GLACIERS OF THE ROCKIES

COPYRIGHT © 2018 GARRETT FISHER. ALL RIGHTS RESERVED.

ISBN: 1985089246
ISBN-13: 978-1985089242

PUBLISHED BY TENMILE PUBLISHING LLC - ALPINE, WY
WEBSITE & BLOG: GARRETTFISHER.ME

FRONT COVER: HELEN GLACIER - WIND RIVER RANGE, WY. **ABOVE:** WEASEL COLLAR GLACIER - GLACIER NP, MT. **REAR COVER:** HARRISON GLACIER - GLACIER NP.

Table of Contents

Flying the Glaciers — 3

Front Range, Colorado — 5

Teton Range, Wyoming — 13

Absarokas & Big Horn Range, Wyoming — 38

Wind River Range, Wyoming — 42

Absaroka Mountains, Montana — 69

Crazy Mountains, Montana — 83

Mission Mountains, Montana — 87

Swan Range, Montana — 91

Flathead Range, Montana — 96

Glacier National Park, Montana — 101

Above: Icefield Pass, CO.

Flying the Glaciers

Since the first time I laid eyes on a photograph of alpine tundra and snowcapped peaks as a kid, I felt an irresistible attraction that can only now be described on a level that borders on the spiritual. There is something incredibly powerful about reaching mountain heights that protrude so far into an inhospitable atmosphere that a quantity of persistent cold and snow exceeds that which summer sun can melt, resulting in a literal river of ice that grinds and crushes everything in its path, forming very severe mountain ranges that are left behind. While on one hand a glacier can represent an environment where humans cannot live permanently, on another it represents the beginning of a cycle of life, where snowmelt can form the basis of our major river systems, which support agriculture and therefore life as we know it. I cannot help but feel a part of this entire paradigm when viewing a glacier.

I can trace the genesis of this project almost 20 years ago, in my late teen years, when a rather studious friend of mine had come across an initial study that the glaciers in Glacier National Park were forecast to be entirely melted by 2030. I remember being dumbstruck by the news, as I had spent my entire life up until that point living in Upstate New York, had never once travelled to the West, and yet still found the whole affair to be disturbing. The West represented to me, in an exaggerated form, what it represents to many in America: a sort of spiritual destination, an inevitability, a purity, freedom, and untouched wilderness. It was a form of the "real" America that I felt a certain pull toward, and in hearing that the glaciers on those snowcapped peaks would disappear, I felt that a significant piece of this pure Americana would be gone forever, and I needed to see them all before it did.

Fast forward many years into the future, and I found myself living in Wyoming, not too far from Yellowstone, with an antique airplane. A story in and of itself, the 1949 Piper PA-11 was restored by my grandfather and happens to be the airplane that I took my first solo flight in and engaged in most of my initial flight training. It is the earliest in its model lineup of bush aircraft, setting the stage for the Piper PA-18, known as the Super Cub, which is the de facto bush plane of Alaska to this day. It is a slow and simple aircraft, cruising at gentle speeds with rugged ground characteristics, good climb rates, and relatively safe handling, meaning that it's a fantastic airplane for those that wish to get back to what is called "stick and rudder" basics, free of excessive electronics and other contrivances. While it can be a lot of fun, there are limitations when flying slow, without heat, and without modern conveniences, and I have not let those limitations get in my way. The airplane became a perfect aerial photography platform such that I came to Wyoming with a long list of things to see, photograph, and write about.

My initial knowledge of glaciers was limited. While I was aware of the glaciers in Montana and held the disposition that the West was a land of verdant wintry pines akin to Alaska or Canada, I also knew that Colorado wasn't known for its glaciers, and that lack of water is a big issue. I guess I was like the typical American that hadn't traveled through the West, imagining the whole thing to be one giant perfect tourist brochure, then realizing that there are deserts, dry areas, and the whole place starts on fire…. but there's lots of snow too.

Some of these prejudices were clarified in my time flying around Colorado, where I was able to see and photograph the peaks over 14,000 feet, many in winter, showing an incredible amount of snow. I also saw how dry Colorado could be, and how these two realities could co-exist. Despite towering to such immense heights, the snow did melt off each summer, with rare exception, and despite some occasional snowfalls in the summer, the entirety of Colorado would become snow free for the summer eventually, which gave rise to the nagging question: how does so much snow, so much dryness, a lack of glaciers here, and yet glaciers in Montana work? There had to be a link to the whole thing, and there was no better place to be than in the middle of Montana and Colorado in the state of Wyoming, to be incentivized to figure it out.

Naturally, a period of intense research followed, where I was determined to find out exactly where the glaciers were. Rumors abounded, as did information from the airplane that initially proved to be false: abundant snow features that lingered for a good portion of the year in peaks around Colorado, only to melt off, though not all of them. If it was a good year, some patches of snow would remain into next season, though that did not make them a glacier. I found out in my extensive digging that "perennial icefields" are not the same as glaciers, as the difference primarily has to do with the fact that glaciers are rivers of ice and snow and literally move down the mountain, whereas stagnant snowpack that sits over the summer is different.

There was then the matter of areas that recently had glaciers, or had glaciers in the ancient past. Up until 1860 or so, many glaciers had recently grown around the world, though the extent was nowhere near what ancient history had to say. Flying around Colorado, it was clearly evident that there were large sections of the Rockies that had glaciers at some point a very long time ago.

Research proved my suppositions to be true, that large ancient glaciers existed in a significantly larger portion of the Rockies than most would expect, and that timberline used to be far lower. However, the ancient past, while informative and interesting, is not the present, nor is it our recent past. Hidden amongst evidence of ancient glacial beds, annual snowfall, and scattered perennial icefields lies present day glaciers, and these are the ones I was looking for.

There was an added wrinkle: many of these present-day glaciers have severely receded or melted since 1860. Some have names; many do not. Some named glaciers no longer exist, whereas other glaciers without a name still exist to this day. That set off an even deeper amount of research, where I had to answer the question: if I wanted to hike to a real, live glacier today, in the American Rockies, where would I find them? After great amounts of work, I settled on a list of mountain ranges, ranging from northern Colorado, through Wyoming, and terminating in Montana at Glacier National Park on the Canadian border. What I did not find, to my surprise, was any remaining glaciers in other Rocky Mountain states: Utah and Idaho, even though in many living person's lifetimes, a few remnants did exist for a period. For the purposes of this book, I do not include the Sierras of California or the Cascades of Washington and Oregon as part of the glaciers in the Rockies, as they are separate mountain ranges. Rest assured, the Pacific Northwest has an enviable number of glaciers to chase, though I reserve that for another day.

I was surprised to find that Colorado had more glaciers than I expected, though they were small due to being so far south. Montana had less than I expected, with Glacier National Park being the southern terminus of a glacial climate that protrudes well into the Canadian Rockies. While a beautiful park in and of itself, if one were to compare it to what lies 300 miles north, Canadians would not be likely to notice Montana. The real surprise was that the largest glaciers in the US Rockies are not in Montana at all; they are in the State of Wyoming, in the Wind River Range, which lies in the middle of the state. Surrounded by open range that is substantially desert, the Wind River Range is a designated wilderness area, meaning that it is only accessible on foot. That makes it utterly severe to access these glaciers, which excludes almost the entire population, save for a few rather rugged hikers.

Once the glaciers were located on a basic level on a map, there was a matter of photographing them. I made a decision early on that I preferred to see to them during annual snowmelt, as while glacial features are interesting under fresh snow, they are more interesting when cyclical snowfall melts off and reveals what lies beneath. As previously indicated, a deep pile of snow in Colorado can deceive a visitor into thinking a glacial feature lies beneath, when it is merely seasonal precipitation. I knew from Colorado that to truly achieve complete melting of snowfall, I would likely have to wait until August. While July was theoretically possible due to the prior winter having been rather dry and warm, there is a reality that glaciers tend to be in shadowed, high elevation areas, and they have ice underneath. That means that they are the last to have snow melt for the year, so reality gave way, and August became the target window to get it done.

When it came time to take the flights, the most complicated part is mountain flying. While I could devote an entire book to the subject of flying small planes around mountains, suffice it to say it is a complex task with many dangers to be avoided. Winds can be severe and complex, and they must be well understood to avoid encountering dangerous downdrafts. Since my aircraft is gentle and forgiving, that also translates into having a lack of power. If the winds are blowing down, most of the time my only option is to escape said wind; I cannot overcome it with engine power alone. Thankfully, I had a reasonable amount of mountain flying experience having attacked the fourteeners of Colorado, so I understood what I needed to do.

That didn't make it easier when many of the flights came along. Part of why glaciers exist in some of these places is due to the fact that they are along a border of climate areas. One will note on the map that there are not glaciers in the interior mountain ranges of the Rockies; rather, they tend to cluster at the intersection of the Great Plains, where a lot of meteorological activity is occurring, including wind. In almost every single flight I took for the project, there was some strong winds blowing, unusual for the circumstances, and most of the time related to being so close to the Plains. I found over time that interior mountain ranges, while still being very high in altitude, had a lot less wind than those that abutted downwind open space, which created a vacuum effect and allowed winds to accelerate. Cruising at altitudes up to 15,000 feet, I found myself snaking around towering peaks and doing my best to take loads of pictures, locate glaciers, fly an airplane, and avoid downdrafts. It was unquestionably a memorable experience albeit a difficult one.

I was initially unsure if I would have the ability to complete the glaciers of Montana, as I had limited time in Wyoming before a move to Europe. As is usual, I had underestimated the magnitude of the Colorado and Wyoming glaciers, having pushed the last of them into September, delayed by smoke from seasonal fires and other weather concerns. Exhausted after an eight-hour photography flight to the Big Horn Range, I said to myself "I don't think I can handle the Montana glaciers; it's just too much." The next morning, I woke up to a retro poster promoting Glacier National Park, which was hanging in the master bedroom, and said to myself "I have to do it!" Four days later, I was in the airplane heading north for a massive flying bender all the way to the Canadian border, outrunning smoke coming from the south and incoming early season snows from the north. I was unquestionably glad I undertook the second phase of the project, as by the time a reader picks up this book, the glaciers already do not look the same. Each year they are shrinking in size, and never in our lifetimes will we see what was available when I took these flights. Soon enough, Glacier National Park will have no glaciers, and that will be that.

While the subject is charged with differing views, there is no debate that the glaciers are disappearing. What we may do about it is subject to society's will; however, I find that it is extremely informative to have a look at the subject in question as we decide what may be done. Similar to my "Where the Colorado River is Born" project about the Colorado River watershed, it occurred to me that few lawmakers and citizens have physically laid eyes on these national treasures. We talk about them, have our thoughts about them, and decide accordingly what action we want or do not want to take. However, not many of us have looked at them in person, which is a logical prerequisite to an informed decision. A significant reason why these glaciers are so little viewed has to do with the magnitude required to visit them in person, particularly on the ground. Spread over hundreds of miles and in the shadows of some of the most severe terrain one could find in America, I would be hard pressed to find a single human that has hiked to all of them, as it would take untold years, determination, and incredible skill to pull it off. For that, I am thankful to have access to an airplane, which made all of it possible during the summer season of 2015, which therefore makes it possible to share with the rest of the world.

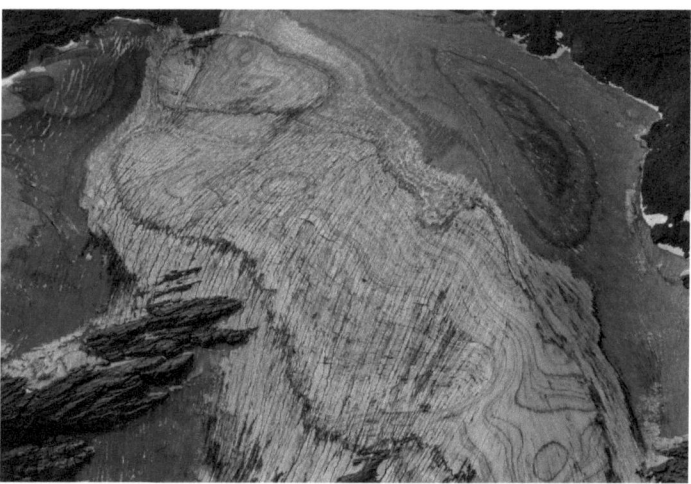

Ahern Glacier, Glacier National Park, MT.

Front Range, Colorado

The reaction from most people when mentioning glaciers in Colorado is a look of surprise, followed by disbelief, with an utterance of "There are glaciers in Colorado?" Unlike their counterparts in Alaska and Antarctica, the glaciers of Colorado are a shadow of what they once were in prior times, though enough remain to meet the criteria.

Remaining glaciers are found in northeastern Colorado, in the Front Range in and around Rocky Mountain National Park. There are plenty of ice fields, perennial snow features, and periodic winters where some snow remains in other parts of the state, and an abundant record of prior glacial activity. Rock glaciers are especially prevalent, which are streams of rock talus that look and sometimes behave like glaciers, sliding down the mountain. Origins vary, inclusive of some rock glaciers having ice beneath them, which lubricates the rocks and allows them to slowly move downward. Many rock glaciers are the results of now expired glaciers, where the snow and ice have melted on the surface, and that which remains helps move rocks that had fallen onto the snow and ice over very long periods of time. There are lots of these to see in Colorado, though they do not feature here.

The Front Range is the windiest mountain range in Colorado, especially when it comes to aviation. As upper level winds crest the mountains, they descend downward and eastward into the Great Plains, picking up speed due to gravity and also the fact that the Plains are uninhibited. Of all of my flying in Colorado, these mountains feature some of the most challenging winds to deal with, and it was no different on this flight, wandering around at 14,000 feet, steering around cloud formations and mountains, and trying to find what remains of our glacial past so far south.

As I have noted previously, the Front Range is another example of many remaining glaciers being found on mountain ranges that form the border of the Intermountain West and the Great Plains. While I have not delved into the subject in a scientific sense, my borderline nerdish infatuation with weather and climatological data has yielded knowledge that the Gulf of Mexico plays a part in these ranges that border the Plains. If a low pressure zone strengthens to the south of any point along these ranges, it arcs warm, moist air south to north from the Gulf, turns it westward over the Plains, where it ascends slowly with terrain. That ascent creates lift and precipitation, which then receives a draw of cold, Canadian air, finally impacting these large mountain ranges with a fully saturated air mass, resulting in tremendous amounts of snow. Some amazing records have been set for snowfall in the United States in the eastern foothills of the Colorado Rockies, a result of this phenomenon. I can only suppose that in the distant past, this kind of weather dumped even more snow, leaving behind the glaciers we can see today.

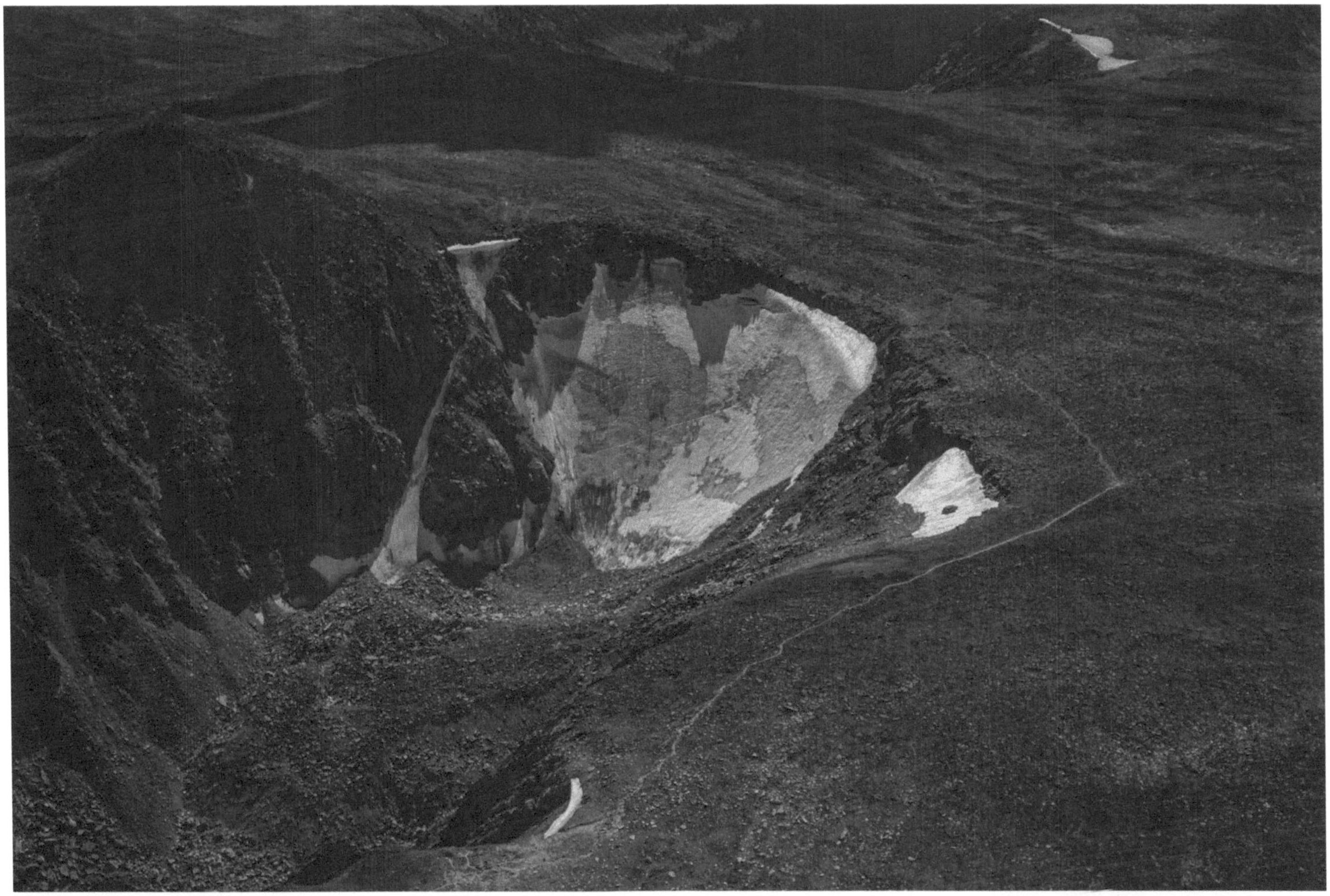

LEFT: *Andrews Glacier.* **ABOVE:** *Tyndall Glacier.* **BELOW:** *Buchanan Pass.*

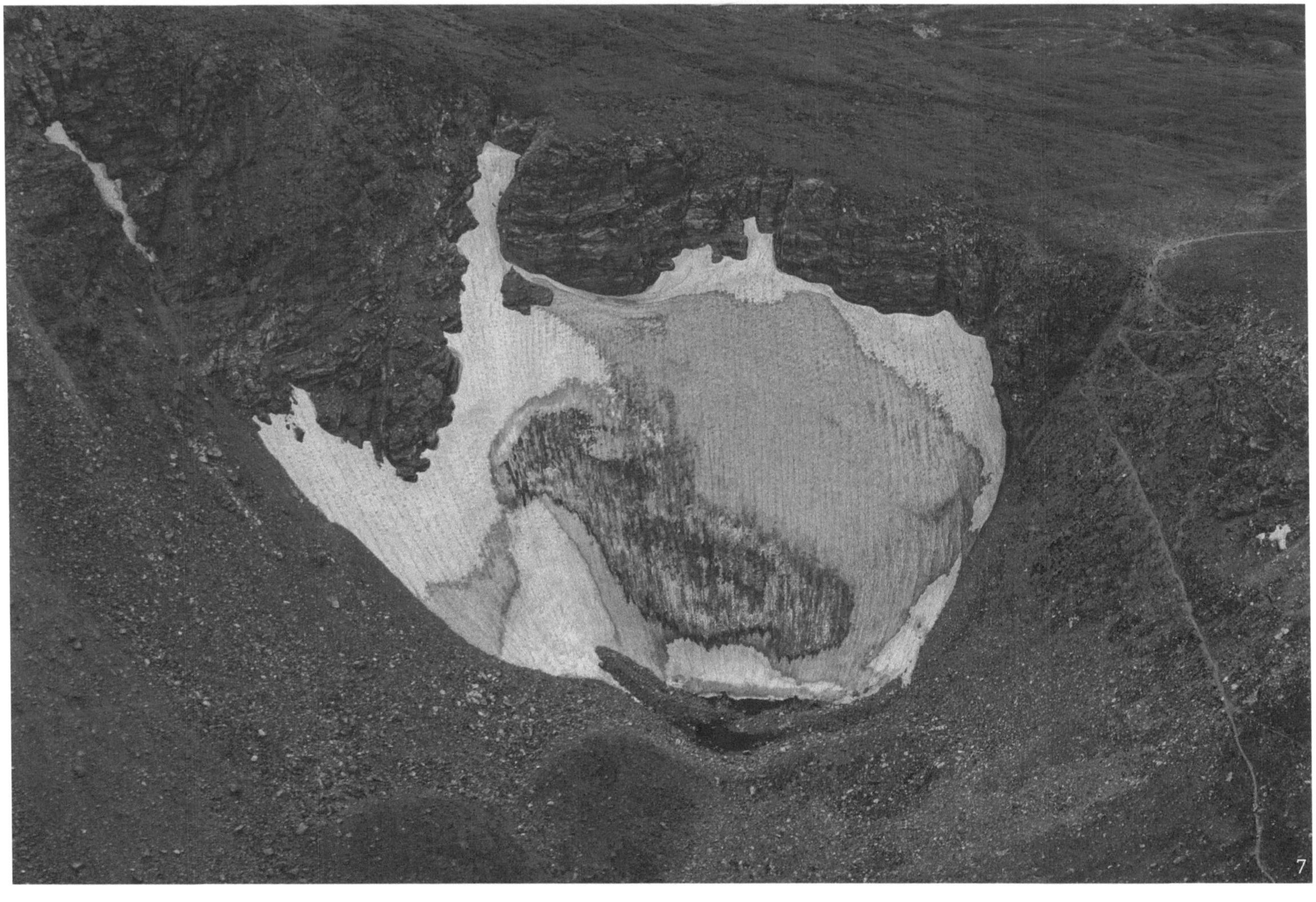

Above: *Saint Vrain Glaciers, Ogalalla Peak (13,077').*

Below: *Arikaree Glacier.*

Above: *Longs Peak (14,259').*

Below: *Moomaw Glacier.*

ABOVE: *Cony Pass.*   **BELOW:** *Ptarmigan Point (12,267') with Tyndall Glacier in rear.*   **RIGHT:** *Arapahoe Glacier.*

ABOVE: *Spectacle Lakes beneath Ypsilon Mountain (13,441').*   **BELOW:** *Rock glacier beneath East Desolation Peak (12,946').*

Teton Range, Wyoming

Prior to the Teton Range, I had not before seen a glacier from the Cub. While I had flown all over Colorado, any glaciers I may have incidentally seen would have been covered under deep snowfall, indistinguishable from the rest of the snowpack. The first time I was able to see them was during a recreational flight to Yellowstone from Colorado, as a part of a "quick detour" during my flight back east after moving from Colorado. I encountered some rather rocky winds while flying along the Teton Range, terrain that towered to rather impressive vertical extremes, unlike anything I had seen before in higher terrain to the south. As I was getting knocked around wondering what kind of nonsense I had gotten myself involved with, I couldn't help but think to myself, "and there are glaciers in there!"

I developed such a fondness for the Teton Range that I devoted the publication of an entire book to flying around the mountain range. Nonetheless, for the purposes of this project, there were roughly 12 named glaciers lurking inside, and I was determined to find them in the right conditions. That meant getting over some fears about the backcountry, vertical terrain, and how aviation mixes with those factors.

The glaciers of the Tetons are glued to the side of very massive terrain, most of the time hiding on the north side of a mountain accessible down a deep canyon, or on the west side, lurking out of glide range to Jackson Hole. Yet another range with some unique weather, the Teton Range features a wide-open fetch devoid of mountains over the Snake River Plain, a product of millions of years of the Yellowstone volcanic hot spot obliterating north-south mountain ranges found over most of the West. That allows Pacific moisture to reach inland with abundance, creating bigger glaciers, and fantastic snow for famous Jackson Hole.

ABOVE: *Falling Ice Glacier, Mt. Moran.*   **BELOW:** *Schoolroom Glacier.*   **RIGHT:** *Triple Glacier, Mt. Moran.*

Left: *Schoolroom Glacier & Grand Teton.* **Above:** *Unnamed glacier, west of Middle Teton.* **Below:** *Snowdrift Lake & unnamed glacier.*

ABOVE: *Petersen Glacier.*

BELOW: *Unnamed glacier beneath Maidenform Peak.*

ABOVE: *Beneath Middle Teton, west side.* **BELOW:** *Unnamed glacier on northwest slope of Veiled Peak.*

LEFT: *Triple Glacier.*   **BELOW:** *Northeast slope of Mt. Moran.*   **ABOVE:** *North slope, Mt. St. John.*

ABOVE: *Unnamed glacier beneath Buck Mountain.* **BELOW:** *Teton Glacier.* **RIGHT:** *Middle Teton Glacier.*

LEFT: *Unnamed glacier on Table Mountain.*   **BELOW:** *Icefloe Lake and related glaciers.*   **ABOVE:** *Petersen Glacier.*

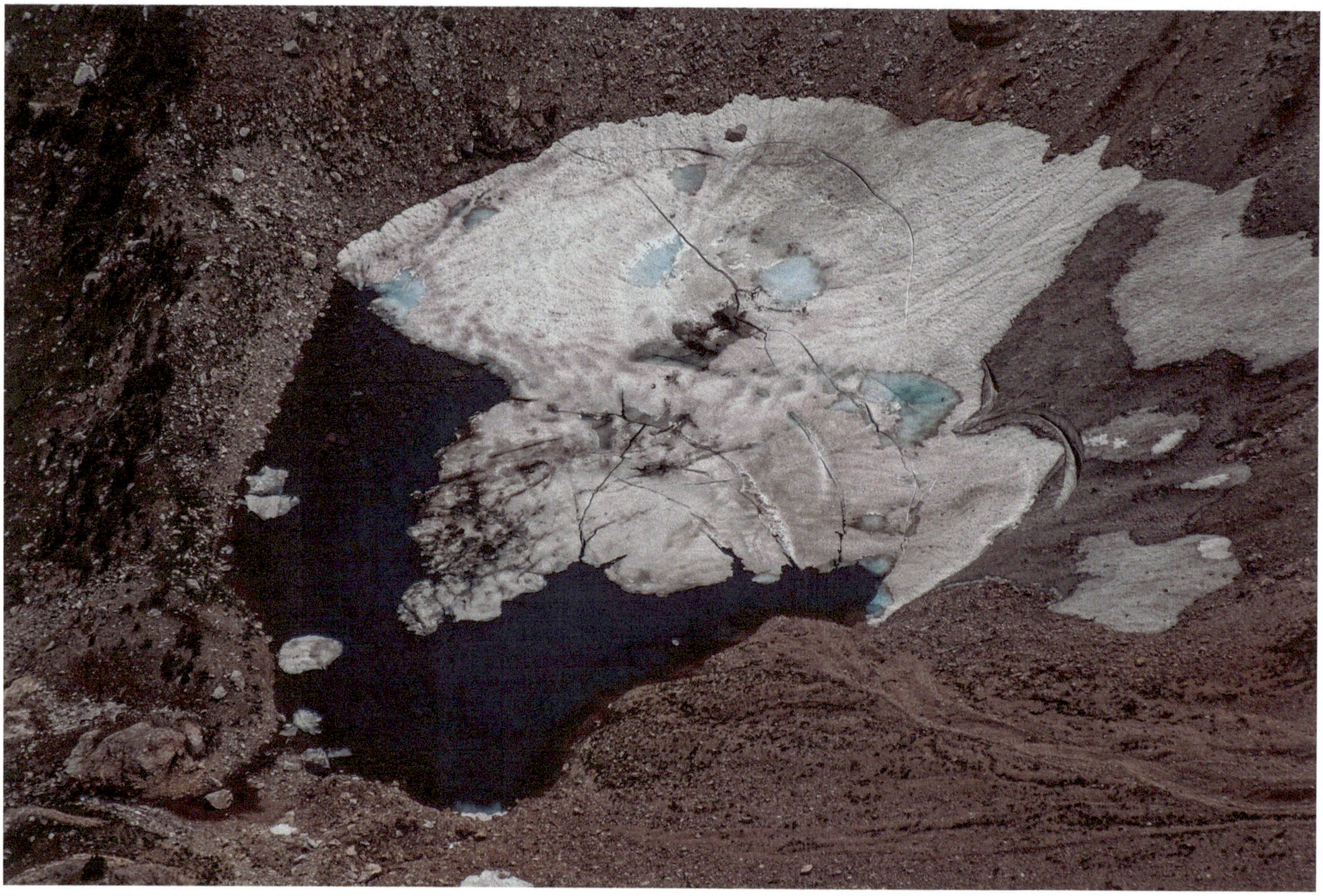

Above: *Glacial remnant beneath Paintbrush Divide.* **Below:** *Icefloe Lake & unnamed glacier.* **Right:** *Schoolroom Glacier.*

LEFT: *Skillet Glacier, Mt. Moran.*   **BELOW:** *West of Mt. St. John.*   **ABOVE:** *Middle Teton with Middle Teton Glacier.*

ABOVE: *Triple Glacier.* **BELOW:** *Remains of glacier beneath South Teton, with ski tracks.* **RIGHT:** *West slope, Grand Teton.*

LEFT: *Falling Ice Glacier, Mt. Moran.*   **BELOW:** *Triple Glacier.*   **ABOVE:** *Triple Glacier.*

ABOVE: *Triple Glacier.*   **BELOW:** *Triple Glacier.*   **RIGHT:** *Glacier just beneath summit of Mt. Moran, northwest slope.*

LEFT: *Glacier beneath Mt. Owen, NE slope.*   **ABOVE:** *Beneath Thor Peak.*   **BELOW:** *Paintbrush Divide.*

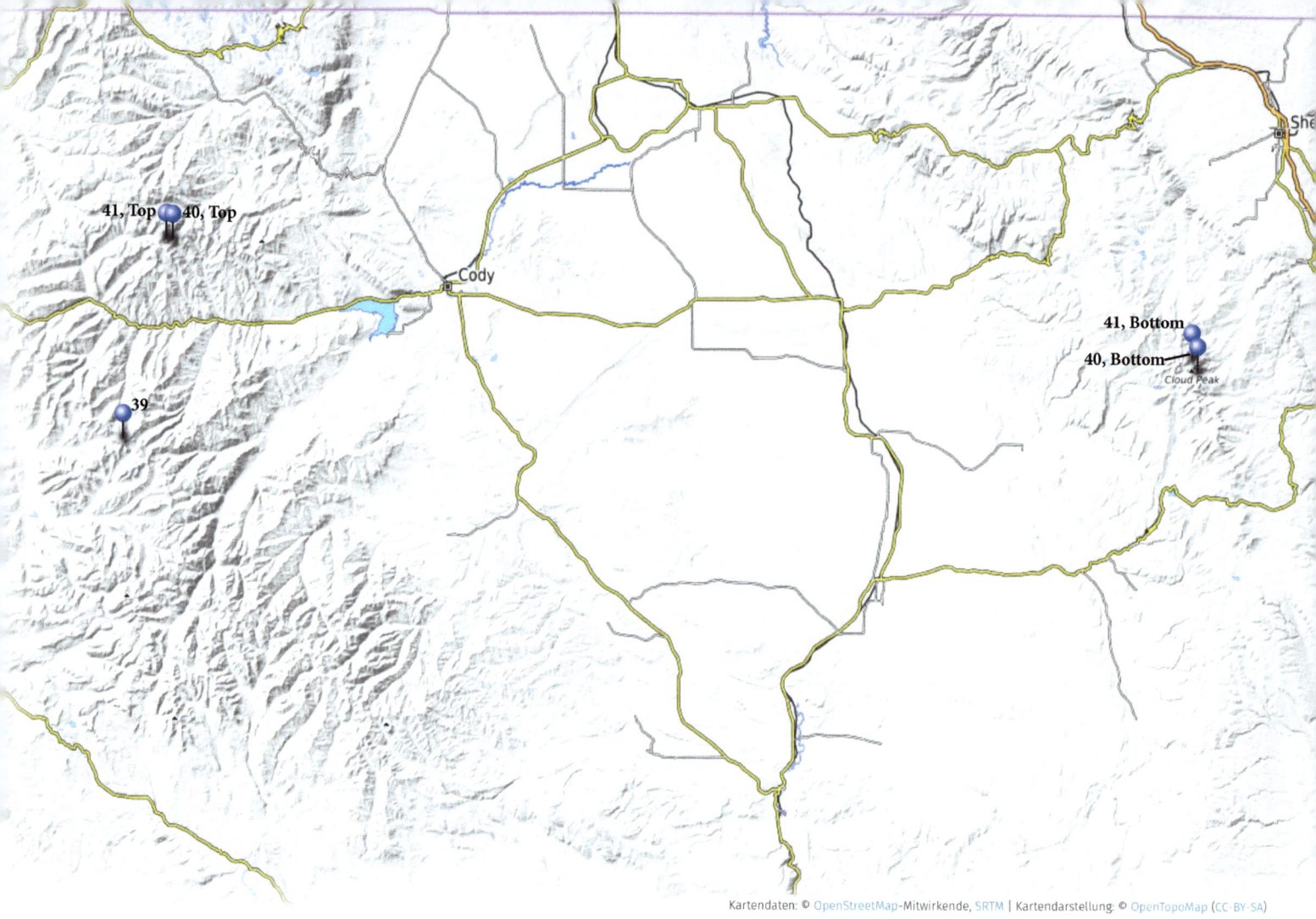

Absarokas & Big Horn Range, Wyoming

My first exposure to the Absarokas and Big Horn Ranges occurred during that initial recreational flight to Yellowstone. After refueling in Montana, it was time to turn around and head east to make the long flight to North Carolina. That meant crossing directly over the eastern entrance of Yellowstone, over the Big Horn Basin and over the Big Horn Range.

I didn't have many initial thoughts of the Absarokas, probably more to do with the weather on that flight, which changed from smoky air and cloudy skies to sunny weather over the Big Horns, which put on quite a beautiful show of texture and color at sunset. I remember thinking of the range as being almost spiritual, as it had a subtle profoundness to it. Dry, immense canyons down low gave way to expanses of pine forest, which terminated in a large crest area. Instead of the typical saw tooth ridge with very little room on top, this mountain range featured plenty of room at 11,000 feet. I was aware of Cloud Peak and its associated glacier and wilderness area, and opted to avoid it, as it looked rather menacing at the time.

Later, as I conquered rather severe terrain when I returned to the West, these places started to seem more doable. I did, however, learn one reality about the Absarokas during a late spring flight to Cody, WY: there is a lot of wind here. The Big Hole and Teton Ranges of Idaho and Wyoming deflect a good portion of prevailing winds, creating a shadow to the east of these ranges in Jackson Hole. Yellowstone, on the other hand, is a high-altitude plateau with the Snake River Range to the west, wide open and devoid of any terrain to block prevailing winds. That creates quite the wind energy, and it exits in two funnels: Togwotee Pass on the south end of the Absarokas, and the valley toward Cody east of Yellowstone.

It is in this valley, during the flight to find the remaining two glaciers in the Wyoming Absarokas, that I broke my groundspeed record in the Cub: 150mph. Winds at 13,000' between Cody and Yellowstone were so strong that I had an 80mph tailwind, something that did not exist in other areas of the flight. As I landed in Cody for fuel, it was rather windy (thought not 80mph), and 10 miles east of Cody over open range, it was negligible.

Right:

Fishhawk Glacier, looking toward Bighorn Basin in mid September.

The Big Horn Range, including Cloud Glacier, was quite a sight to see. Finally entering the Wilderness Area, I noted that the dangers I felt the first time were rather small. A few forest service roads were within reasonable glide and then hiking range in the event of engine failure, which was good enough for me. Winds at this altitude, despite having been on the same day as my speed record, were typical: some speed, though nothing dangerous or turbulent.

These glaciers are the most isolated out of all of them in the US Rockies, as the rest tend to found amongst other groupings. It was at the tail end of a full day of photographing these glaciers that I questioned my stamina to get the Montana glaciers for this project. I was not sure it could be done, though as the rest of the book will show, I couldn't help myself.

ABOVE: *Rock glacier beneath unnamed glacier, Sunlight Peak, Absaroka Mountains.* **BELOW:** *Cloud Peak Glacier, Big Horn Range.*

ABOVE: *Unnamed glacier, Sunlight Peak, Absaroka Mountains.* **BELOW:** *Unnamed glacier beneath Black Tooth Mountain, Big Horn Range.*

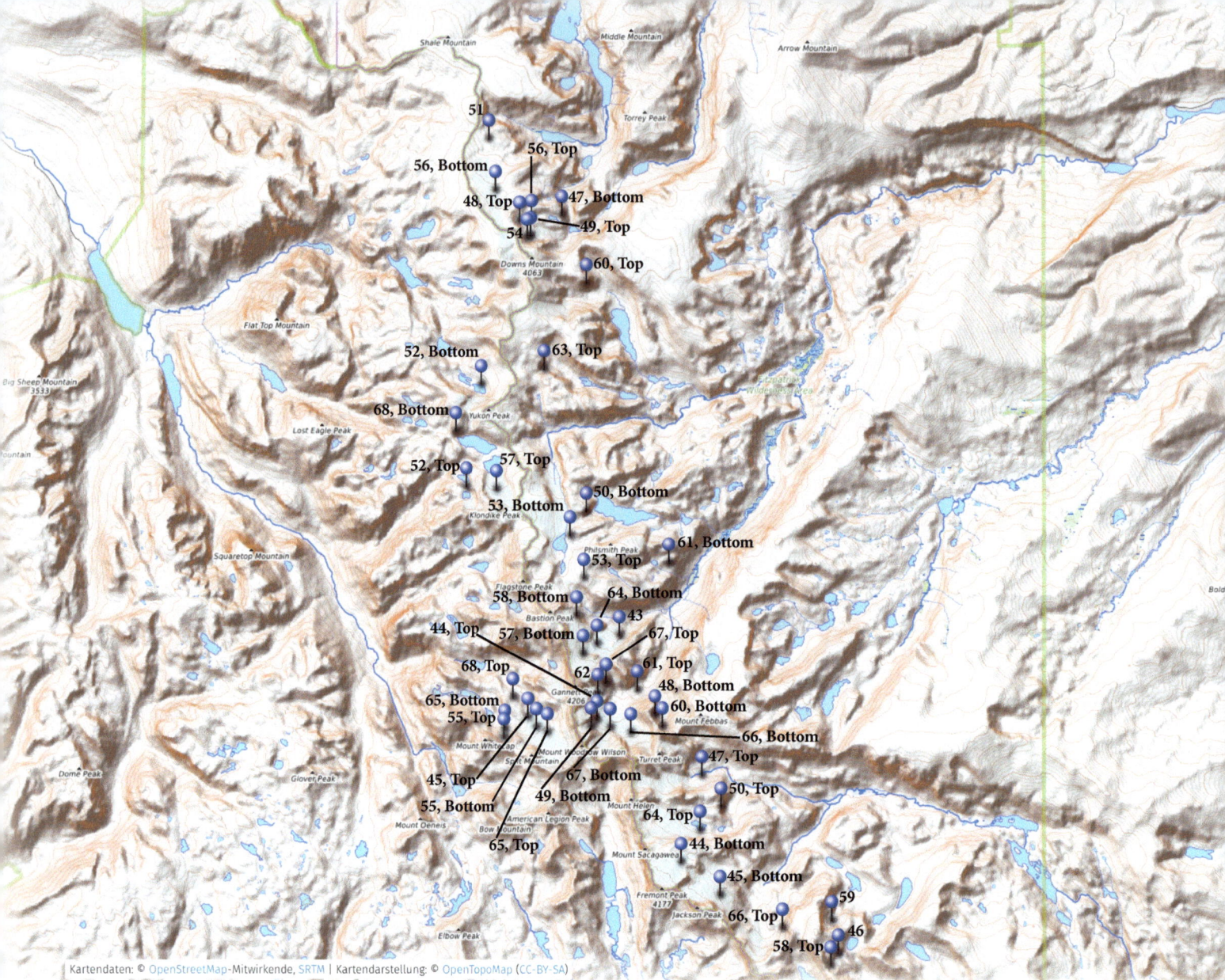

Wind River Range, Wyoming

Wind River Range takes the cake on the glaciers: largest in the US Rockies, most wilderness, and most severe terrain to fly over. These facts were not something I was aware of until doing extensive research, though I was given a clue on that fateful first recreational flight into Wyoming before moving. The mountain range had been a bit attractive as seen from Google Maps on satellite view, seeming to be a much more pronounced geographic feature that required exploration. As I headed north on that flight toward Jackson Hole, I skirted the southwestern corner of the mountains at 12,500 feet, noting that the terrain I found was far more severe than what existed in Colorado. Accordingly, I didn't dive right into the middle, heading northwest due to fuel and time and leaving it at that.

The moment I moved to Wyoming, I knew I would be exploring these mountains, even before I settled on the glacier project. Not too long thereafter, I took a commercial flight out of Jackson, and the trajectory took us along the mountains during climb out, showing them from a very distant view. What I saw showed me how little Google Maps was revealing: terrain was extremely vertical and severe, unlike anything I had yet experienced, which then just gave me a bunch of fear.

Alas, fear usually just translates into a delay while I dig for more information, eventually resigning myself to the fact that the engine usually keeps running, and therefore I don't have to worry about landing on a cheese grater and then becoming grizzly bear food. I took an initial flight along the northwestern corner, and almost couldn't believe what I saw. It may as well have been straight out of Alaska or the Alps: towering terrain with glaciers and spires of incredible rock. While cautious, I was in a way drawn in like a moth to a flame.

Glaciers here are mostly on the eastern side of the range, with some on the west. Some are measured in miles of length, and look like something out of another region, true rivers of ice and snow leaving evidence of centuries of snowfall in their wake. A few perspectives showed small glaciers almost calving into a few lakes.

It was after landing and getting a chance to dig into the details of what I photographed to understand how big some of these glaciers are. Aerial perspective is so distant that things can look much friendlier than they are, and what I saw was rock falls, large crevasses, and a rather significant mass of frozen water that represents a perspective in time, snowfall frozen from before America existed as a nation, that to this day waters some of our agriculture downstream. These glaciers are more than just nominally a glacier by definition, they are big enough to be part of a larger ecosystem, yet far enough away that most do not know they exist.

Right:

Gannett Glacier, with Gannett Peak (13,809'), the highest peak in WY.

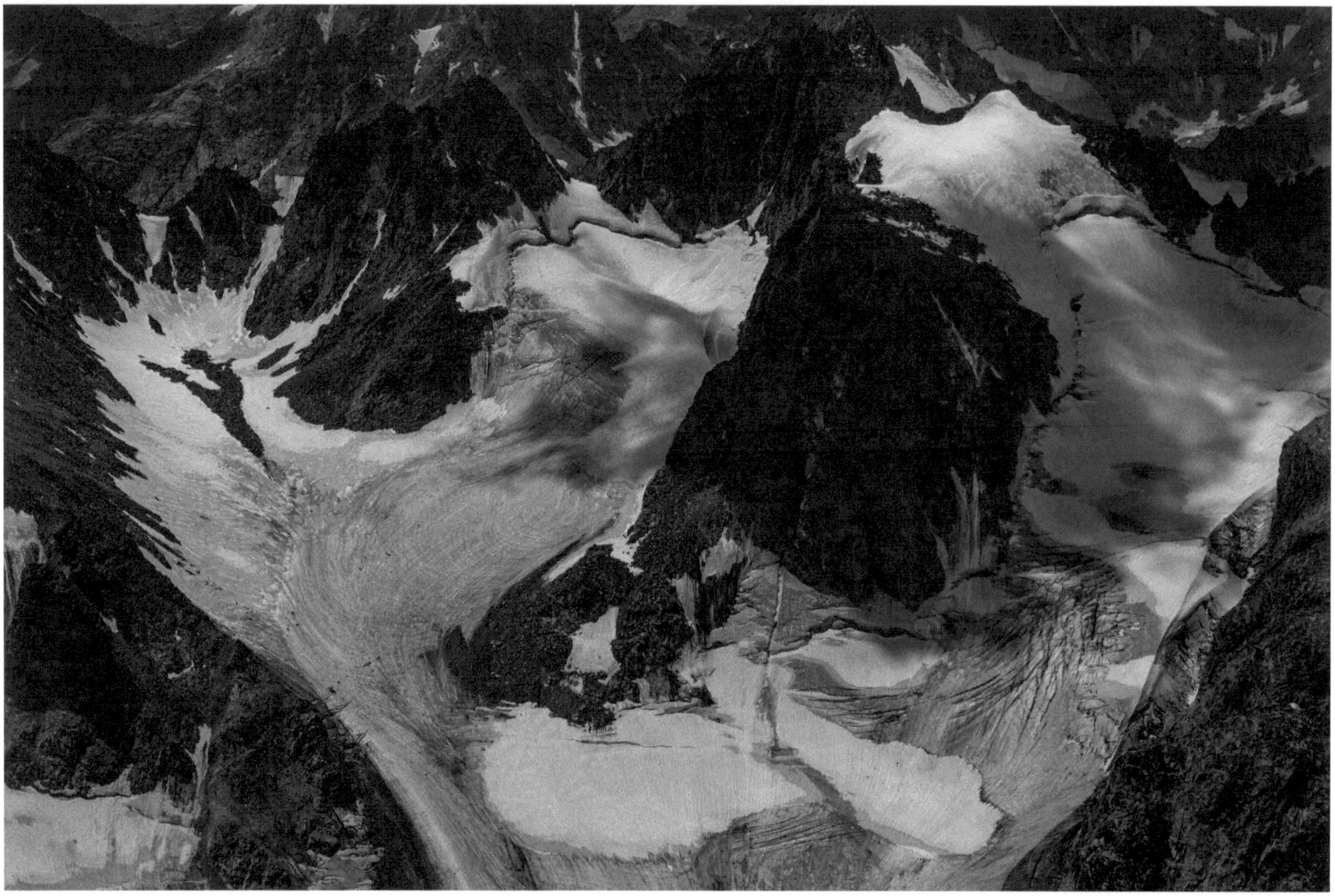

ABOVE: *Dinwoody Glacier.*

BELOW: *Upper Fremont Glacier.*

ABOVE: *Mammoth Glacier.*

BELOW: *Bull Lake Glacier.*

LEFT: *Glaciers near Brown Cliffs.* **ABOVE:** *Helen Glacier with Sacagawea & Fremont Glaciers in rear.* **BELOW:** *East Torrey Glacier.*

ABOVE: *East Torrey Glacier.*

BELOW: *Dinwoody Glacier.*

ABOVE: *East Torrey Glacier.*

BELOW: *Dinwoody Glacier.*

ABOVE: *Sacagawea Glacier.* **BELOW:** *Grasshopper Glacier.* **RIGHT:** *Continental Glacier.*

ABOVE: *J Glacier.*

BELOW: *Connie Glacier.*

ABOVE & BELOW: *Klondike (left) and Grasshopper (right) Glaciers.*

LEFT: *East Torrey Glacier.*   **ABOVE:** *Baby Glacier.*   **BELOW:** *Mammoth Glacier.*

ABOVE: *East Torrey Glacier.*

BELOW: *Continental Glacier.*

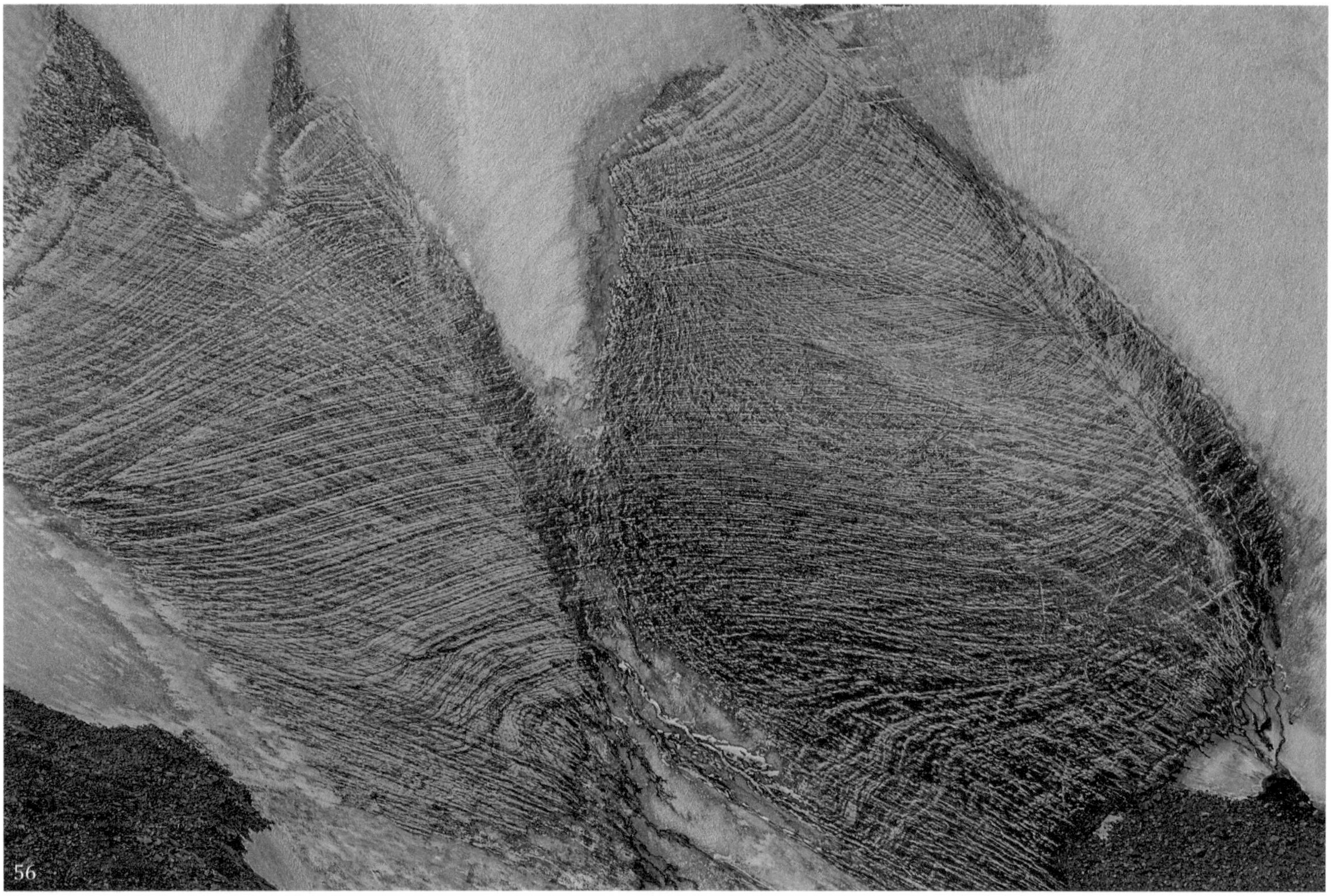

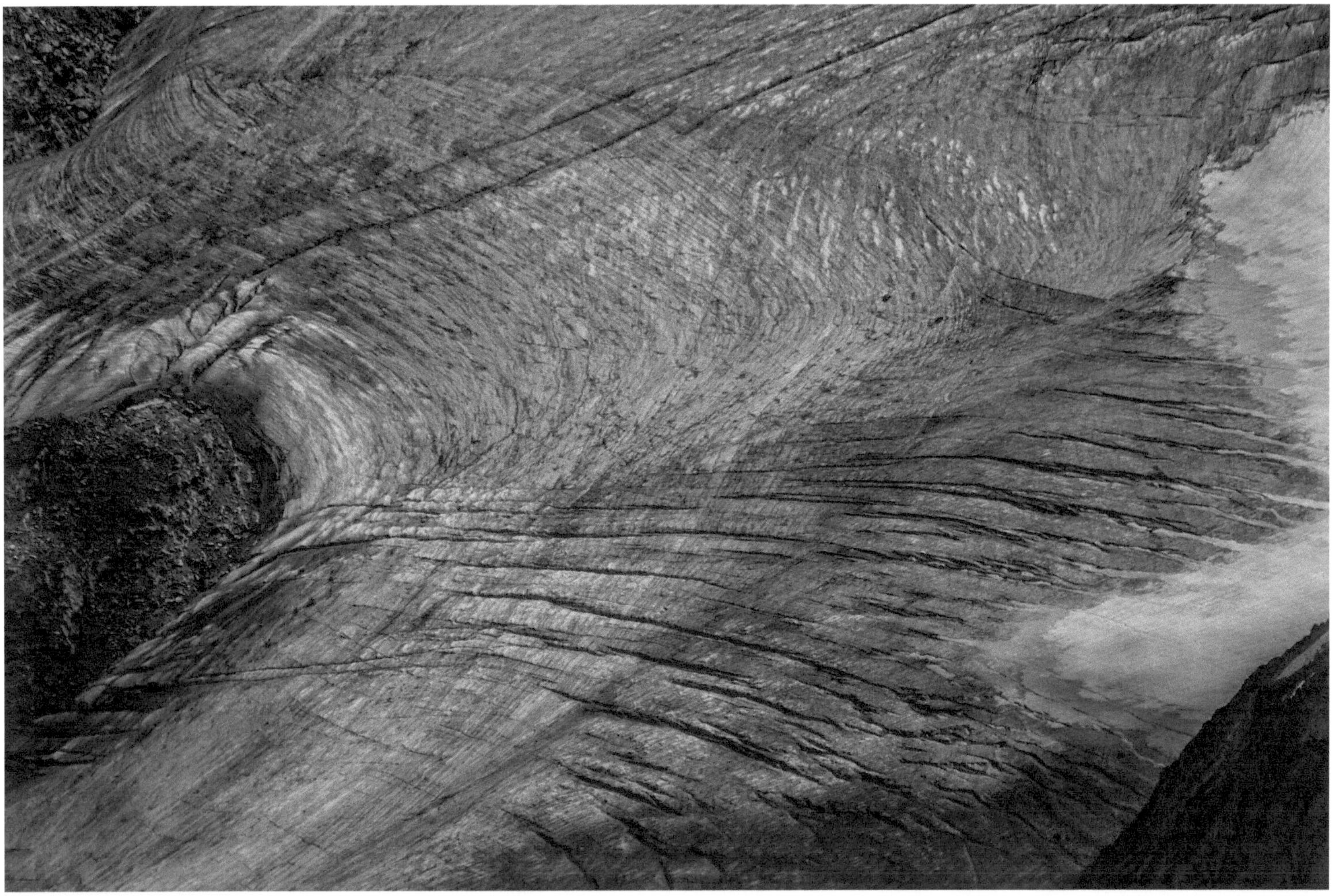

ABOVE: *Sourdough Glacier.*

BELOW: *Gannett Glacier.*

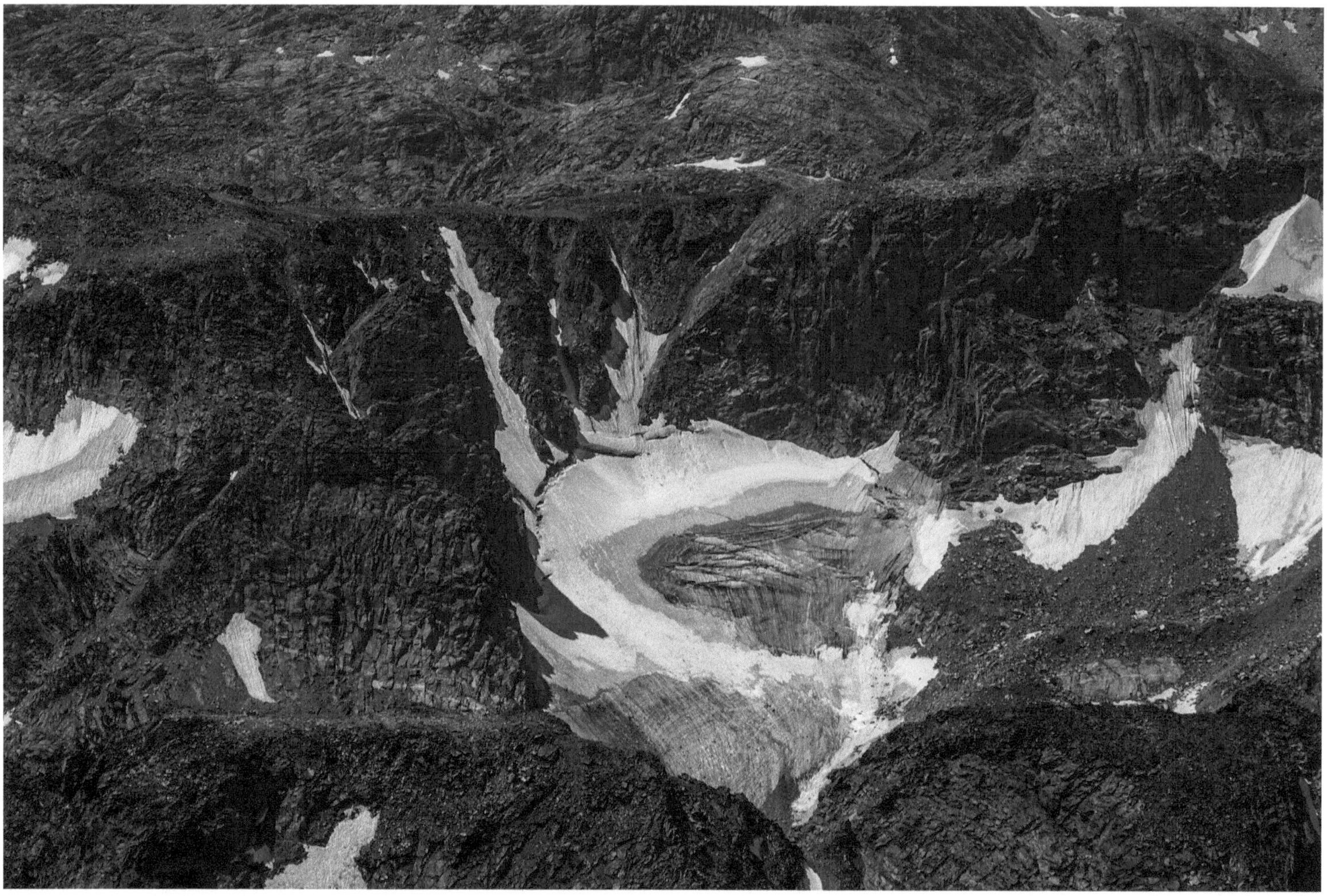

ABOVE: *Unnamed glacier near Alpine Lakes.* **BELOW:** *Gannett Glacier.* **RIGHT:** *Unnamed glacier near Knife Point Glacier.*

ABOVE: *Downs Glacier.*

BELOW: *Dinwoody Glacier.*

ABOVE: *Dinwoody Glacier (left), Gooseneck Glacier (right).*   **BELOW:** *Gannett Glacier.*

LEFT: *Gooseneck Glacier.* **ABOVE:** *Unnamed glacier NE of Yukon Peak.* **BELOW:** *Unnamed glacier.*

ABOVE: *Sacagawea Glacier.*

BELOW: *Gannett Glacier.*

ABOVE: *Mammoth Glacier.*

BELOW: *Baby Glacier.*

ABOVE: *Knife Point Glacier.*

BELOW: *Dinwoody Glacier.*

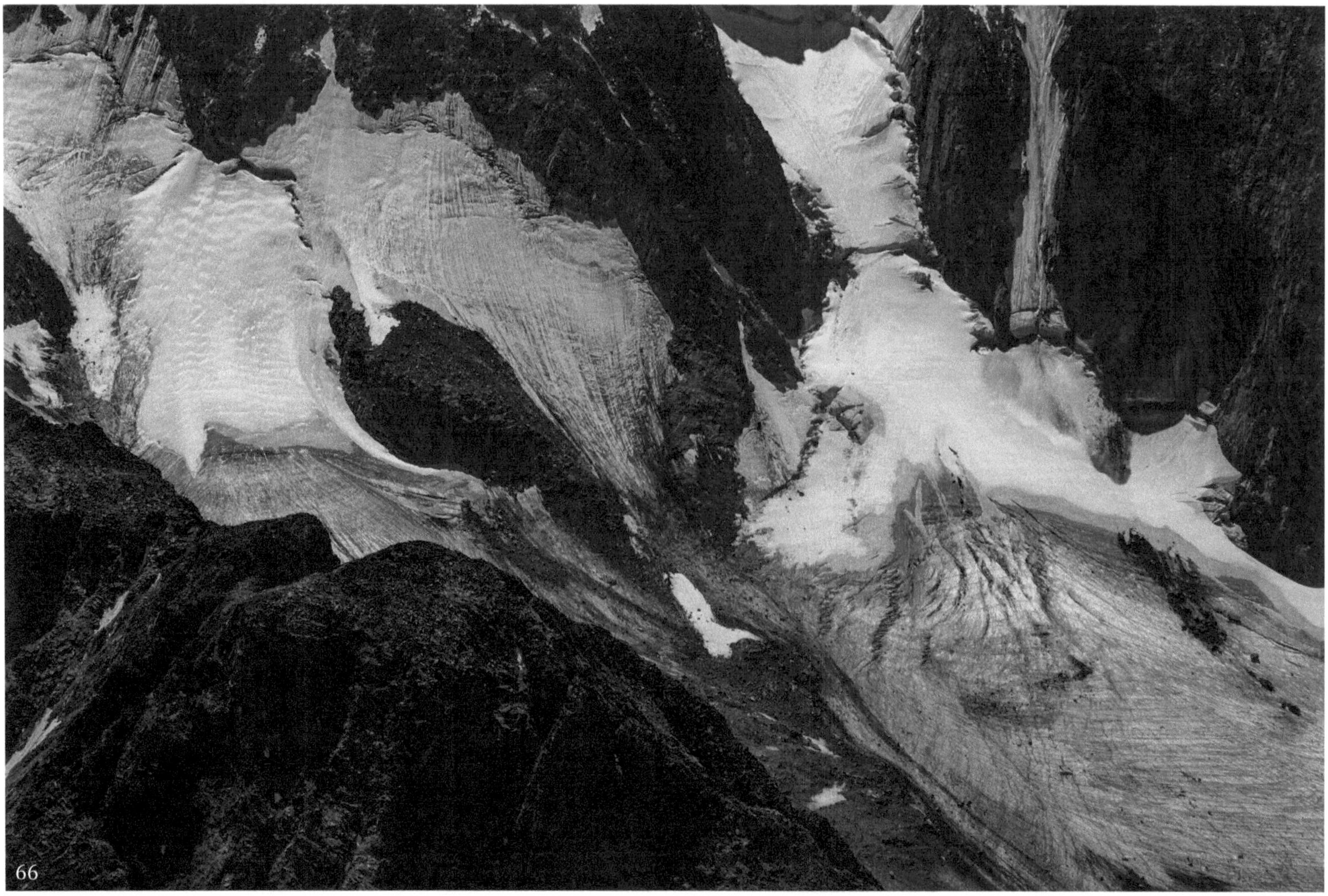

ABOVE: *Dinwoody Glacier (left), Gooseneck Glacier (right).*　　**BELOW:** *Dinwoody Glacier.*

ABOVE: *Mammoth Glacier.*

BELOW: *Sourdough Glacier.*

Absaroka Mountains, Montana

The Absaroka Mountains of Montana are a continuation of the mountain range by the same name in Wyoming, as they cross the border from the south and arc to the northwest. While the mountains in Wyoming are drier and surrounded by other terrain, this range in Montana begins to behave differently on this side of the line. Winds are quite strong, as the Absaroka Mountains become the final range before giving way to the Great Plains. In Wyoming, this is not the case as the Big Horn Range to the east serves this function. By having nothing to the east, winds tend to accelerate, and mountain waves can grow with great ferocity.

The Beartooth Range, a subrange within the Absarokas, features severe mountain waves, which can make flying deadly in some circumstances, whereas it may merely be dangerous, or not a problem at all on the same day in other places. Further to the northwest, at the terminus of the Absarokas in Livingston, MT, winds are so strong that the county airport features an astonishing four runways, pointing in eight different directions to accommodate some of the absurd winds that blow here.

The highest peak in Montana, Granite Peak at 12,807', is found here in the Absaroka Range, along with a large number of small glaciers. Contrary to what most would assume, Glacier National Park's highest summit is only 10,479', despite having a far larger glacial mass than these higher mountains in the Absarokas. I found it interesting that, from a visual perspective, I was reminded most of Colorado when flying in these ranges, including fire-scarred foothills to the east, verdant forests, and rocky summits typical of their cousins further south.

These mountains are also unique in that they are the only mountains on the lee side of Yellowstone that have the Great Plains to the east. As winds roar down the Snake River Plain, reaching the heights of Yellowstone's relatively flat plateau, they must cross these tall peaks before crashing down into the Plains and continuing their eastward march, at speeds typical of the wide-open prairie. While in flight I certainly experienced such realities, and can only wonder to what extent snowfall in centuries and millenniums past was influenced by such weather patterns that remain today.

LEFT: *Castle Rock Glacier.*   **ABOVE:** *Wolf Glacier.*   **BELOW:** *Glacial remnants beneath Granite Peak (12,807'), highest peak in MT.*

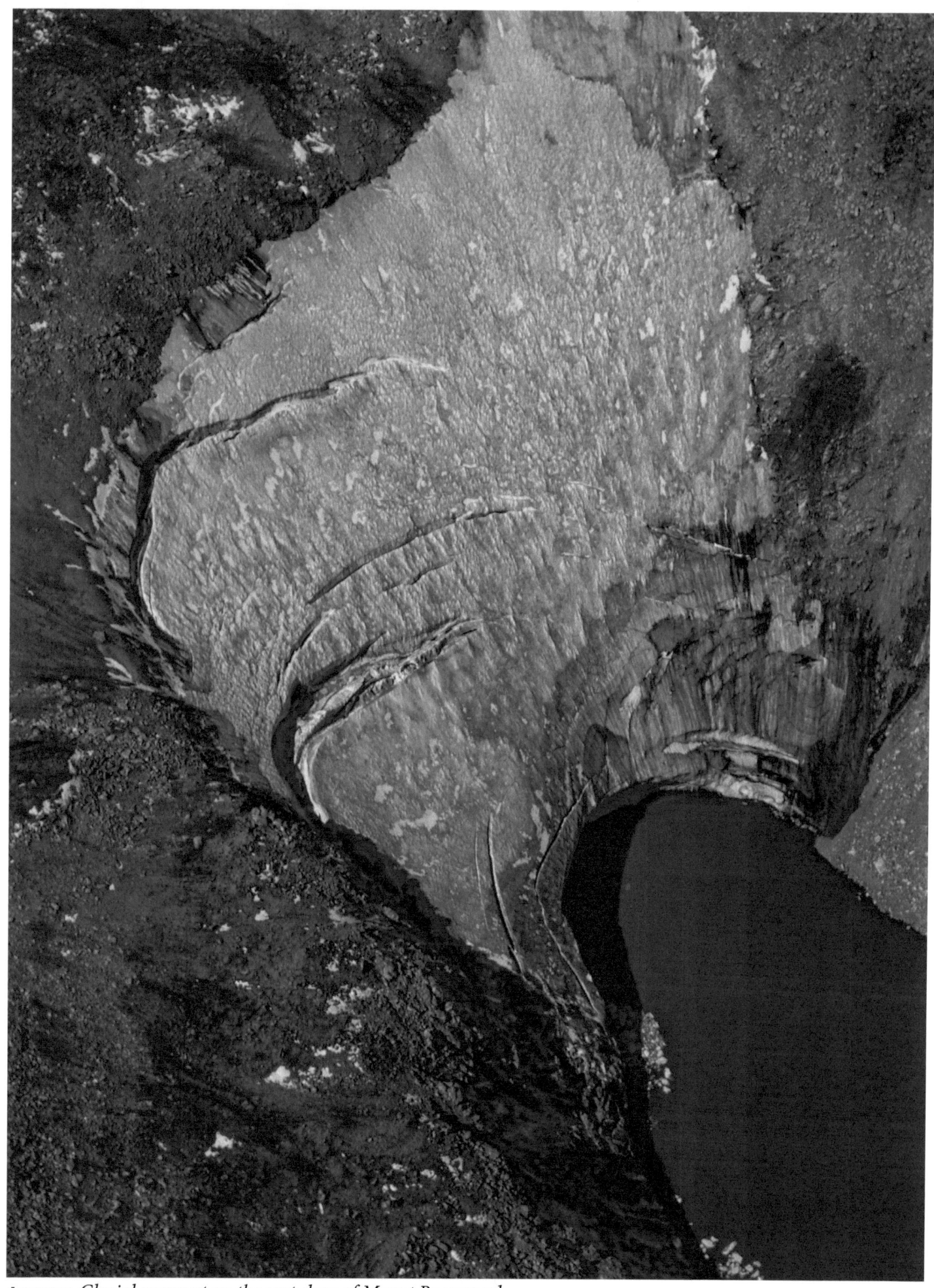
ABOVE: *Glacial remnant on the east slope of Mount Rearguard.*

Above: *Sundance Glacier.*

ABOVE: *Unnamed glaciers beneath Mt. Wise.* **BELOW:** *Grasshopper Glacier.* **RIGHT:** *Unnamed glacier beneath Mt. Peal.*

ABOVE: *Hidden Glacier.*

Above: *Unnamed glacier beneath Big Mountain.*

ABOVE: *Snowbank Glacier with top of Castle Rock Glacier in foreground.*

ABOVE: *Hopper Glacier with series of unnamed glaciers.*

LEFT: *Hopper Glacier*   **ABOVE:** *Wolf Glacier.*   **BELOW:** *Snowbank Glacier.*

Crazy Mountains, Montana

Things begin to change in the Crazy Mountains. A small range found in visual line of sight to the north of the Absarokas, it literally is an island in the sky, surrounded on all sides by prairie. While there are resemblances to the Rockies as I was used to them further south, they begin to take on a different topography here, with steeper mountainsides, and a compressed change in ecology. Literally in one image I can show remnants of a glacier, alpine tundra, high altitude pine, and Great Plains, all in a short lateral and horizontal distance.

The highest peak is only 11,214', yet there are visible ghosts of glaciers past, and these mountains are no more than 40 miles north of glaciers at higher altitudes in the Absarokas. In the present day, climate begins to transition from a purely Intermountain West influence to a modified Pacific climate, found in northern Montana and in the Canadian Rockies. Featuring shorter terrain, higher moisture amounts, lowered timberline, and a greener, wetter appearance, the Crazies offer the first clue of what is to come further north in the same state. In a bit of an irony, the mountains here are higher than Glacier National Park.

Left:

Grasshopper Glacier, Absaroka Range, MT.

On the day in question, I found winds in the Crazy Mountains to be…not crazy at all. Flying was relatively by the book and predictable, until I crossed the Yellowstone River and entered the Absarokas later on. Winds went from predictable to turbulent and aggressive, with changing wind directions, even though I could see both mountain ranges from each other.

ABOVE: *Rock Lake.* **BELOW:** *Glacier east of Iddings Peak.* **RIGHT:** *Glacial remnants to Great Plains - eastern Crazy Mtns slope.*

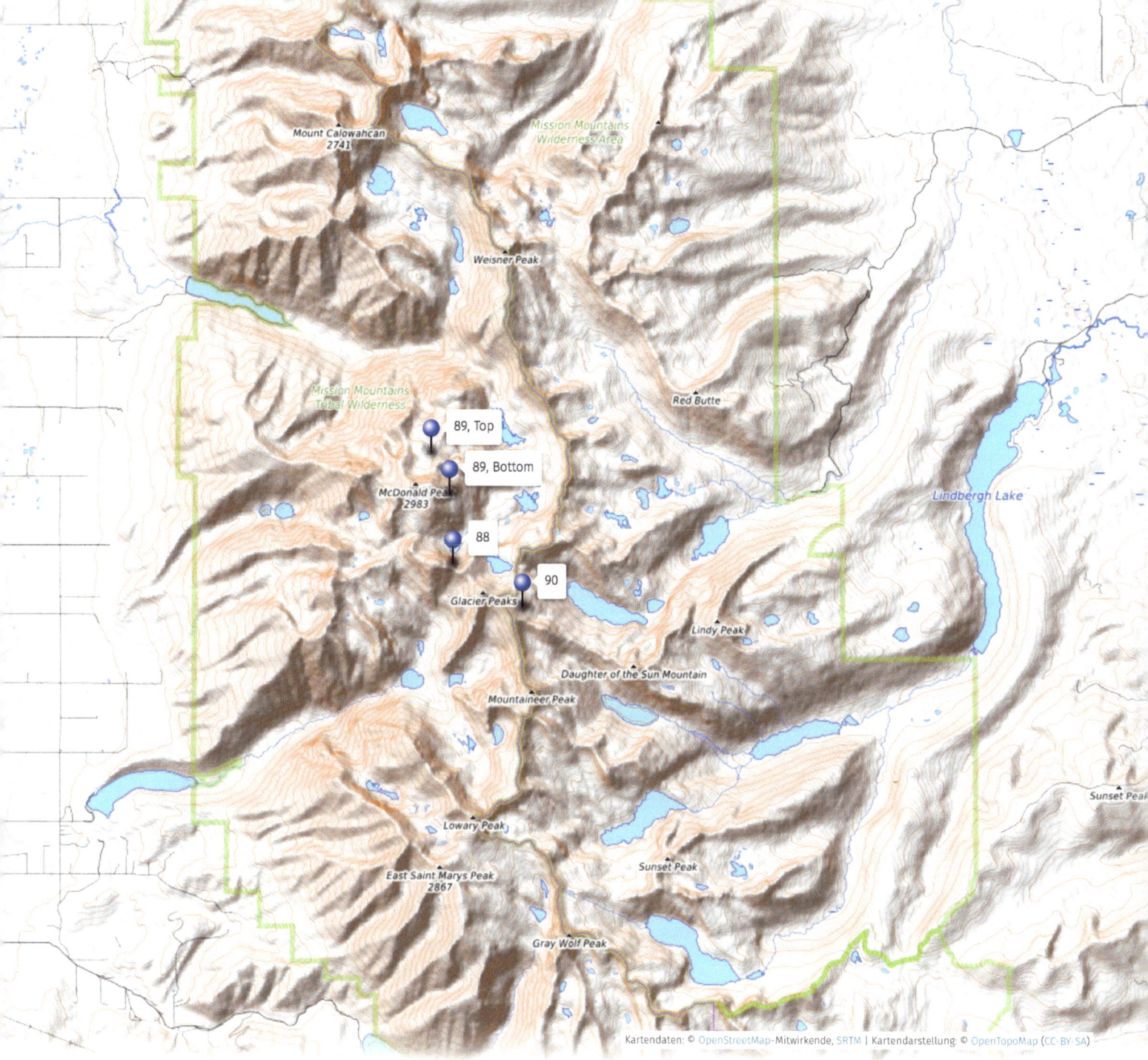

MISSION MOUNTAINS, MONTANA

Between the Crazy and Mission Mountains, terrain is much lower than in other sections of mountainous Montana, meaning that by the time one heads north enough to reach the Mission Mountains, climate and topography have really begun to change. With its highest peak at merely 9,820', well below timberline in Wyoming and Colorado, there are glaciers here, and they are notably different.

Mountain ranges here have a similar saw tooth appearance as further south; however, ranges are much narrower, and compressed in height. Six thousand feet below the Mission Mountains and behind them in some of my photographs, one can see working ranches with habitable four-season climate, yet glaciers in the foreground. Unlike their cousins in Colorado and some of the drier ranges of Wyoming, these glaciers stand a bit prouder. Instead of hiding in recessed shadows, clinging for life until they melt, they are visible in the open, obviously much earlier in the cycle of melting.

LEFT:

Rock Glacier & Swamp Lake, Crazy Mountains, MT.

Larch trees also begin to make their appearance within 100 miles of this range. Larch trees have fascinated me for a long time, as they are pine trees that have needles which change color and drop off for the winter, growing back the next spring like a deciduous tree. For the most part, these trees are endemic to colder areas, and I found this to be the case as I wandered further north for this project, finally gazing into Alberta and British Columbia at the 49th parallel: mountains carpeted with yellow-needled pine trees, denser the farther I went.

While it is possible to see from the air that glaciers in drier mountain ranges to the south were recently larger, it is often the case that their extent may not have been that significant in our current geological era. Perhaps in the ancient past they were extremely large. Here in the Mission Mountains, one can see that glaciers recently were much larger than they are now, as evidenced by their scouring marks and lack of any soils or small rocks to infill a former glacier bed.

Left: *Icefloe Lake.*   **Below:** *East slope of McDonald Peak (9,820').*   **Above:** *McDonald Glacier.*

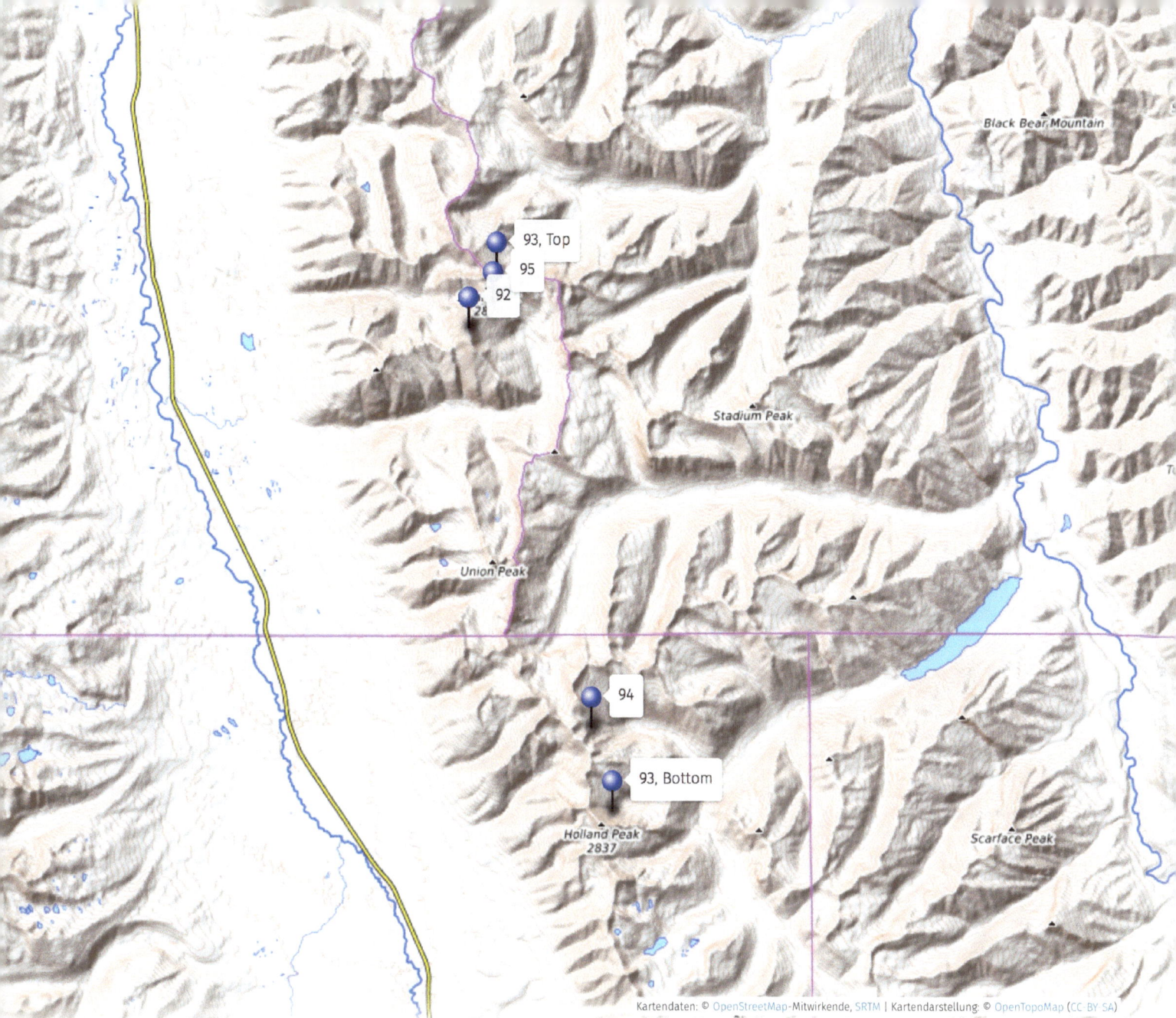

Swan Range, Montana

The Swan Range was a quick hop from the Mission Mountains, though they did give me a little pause. While the glaciated areas of the Missions were a bit wide, meaning that an engine failure meant ditching deep in rugged terrain, they at least had civilization on both sides of the peaks. Whether that civilization is of any use to me in the event of emergency is another story; it is satisfying to have it in visual range, as it provides some basic emotional comfort. The Swan Range only has civilization on one side, to the west, in a rather forested valley, which was admittedly a change from terrain further south. It is unusual to have an inhabited valley completely covered in forest in the Rockies of Colorado, Wyoming, and the southern half of Montana. The reasons likely have to do with original settlement as it relates to snowfall rates, though in any case, it represented a curious demographic change.

To the east of the Swan Range lies the Bob Marshall Wilderness Area. When designated as a "wilderness," an area has special legal provisions which basically prohibit just about everything except walking trails. It was an expansive section of large mountains, complete isolation and rugged terrain all the way to the horizon to the east, and while I couldn't fully identify specific peaks, I knew that the Bob Marshall Wilderness would terminate to the north at Glacier National Park. I could sense that I was approaching some "real" rugged terrain, in a way that I hadn't seen yet further south.

LEFT:

Sunrise Glacier, Mission Mountains, MT.

There were less glaciers here than in the Mission Mountains, though not by a significant order of magnitude. It appears that there were far more permanent ice features in the recent past, though all that remains now are in two areas in a long saw tooth mountain range, with evidence of both recent larger glaciers, and ancient glaciation which was much more significant. Much like the Mission Mountains, the comparison to Wyoming and Colorado was still strange, as vertical distances from valley floor to mountain peak are far less than in neighboring states. The highest peak here is only 9,356 feet, roughly 500 feet lower than the Mission Mountains to the south, yet curiously very low to contain glaciers, as I lived for a year in a valley in Colorado at 9,340 feet.

Left: *South slope, Swan Peak (9,356').* **Below:** *Holland Peak.* **Above:** *Swan Glaciers.*

Above: *Unnamed glacier, Albino Basin.*

ABOVE: *Southeast of the summit of Swan Peak.*

Flathead Range, Montana

It is hard to mention the Flathead Range without mentioning what is going on to the east and north, visible directly behind, which is Glacier National Park. Out of the park boundaries and technically in a separate mountain range, the Flatheads curiously have a maximum elevation of 8,705 feet, while being not too terribly far north of the Swan Range. Mountain ranges in this section of Montana are very close together, without expansive valleys like their cousins further south, yet still retaining geological identity in each one.

Glaciers here are noticeably larger than the Swan Range, though they pale in comparison to the height and magnitude of what lies in nearby Glacier National Park. Another visible feature which lies to the west is the Hungry Horse Reservoir, a rather picturesque and large body of water that occupies the space between the Swan and Flathead Ranges. Glaciers to the east and water to the west is an indication of stronger storm systems and more abundant winter snowfall in these areas, as compared to mountains which visibly appear drier further to the south.

It becomes evident this far north that the Rockies have undergone a complete change in character. While some of it is geology, with differing mountain formations and heights, most of which are steeper and smaller than their cousins to the south, what lies on the surface is far more pronounced. Forests take on a rugged and thick nature, dry valleys disappear, lakes and water become more pronounced. In effect, it begins to look like the Canadian Rockies here, which is a world unto itself in rugged wilderness.

As previously mentioned, larch trees were very evident as they were already in full color in September, showing off a golden hue set against other pine trees, mostly confined to high summits. Grizzly bears inhabit reaches beneath, which was a concern in the event of emergency. The understory, which is the abundance of ground vegetation growing beneath the forest, was in full, explosive autumn color bloom, something I had not seen before in areas to the south. That is an indication of thicker forest vegetation overall, and a sign of further ruggedness.

Right:

Grant Glacier, MT.

It is unmistakable what coexists alongside this moisture and ecological vibrancy: fire and other challenges to the ecosystem. Burn scars exist in these areas, along with vast expanses of forests obliterated by beetle kill, a symptom of increasing winter temperatures creating an opportunity for various beetle species to lay eggs inside mature trees, spreading a deadly fungus and killing entire forests. While these areas are rugged, primal, and spiritual, they are somewhat sensitive to the world around them, and are in close proximity to completely different bio zones, meaning that they are a frontier of changes in our global habitat.

ALL: *Stanton Glacier, with Hungry Horse Reservoir in rear (above).*

ABOVE & BELOW: *Grant Glacier.*

100

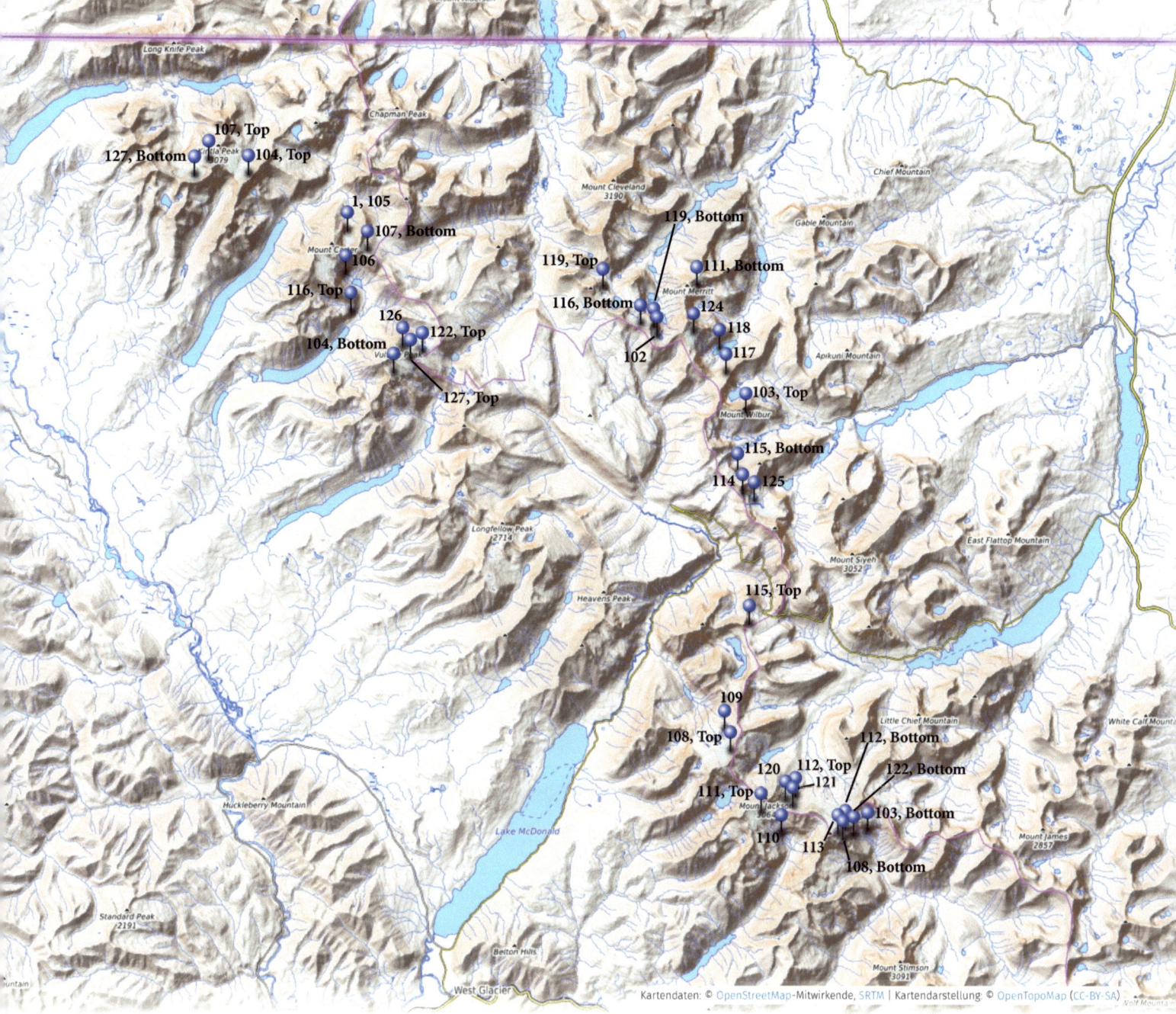

Glacier National Park, Montana

The best is indeed saved for last. While not quite home to the largest glaciers of the American Rockies, the terrain, views, and recurrence of glaciers, with their epic majesty, was by far the best out of all of them. With the highest peak at 10,479 feet, terrain now climbs abruptly on our south-to-north journey through the Rockies, and the glaciers that result are much larger and cover larger areas, as one would expect.

Glaciers here have pronounced crevasses, clear indications of movement, and even some calving off of cliffs, with waterfalls cascading down steep mountainsides, forming the headwaters to rivers that continue out of the mountains and into other areas. There is no question here about technicalities of the definition of a glacier versus an icefield; frozen mass here has a clear magnitude that feels like what visitors would traditionally expect when conjuring images of a glacier in their mind.

While flying over this terrain, I was struck by how little of it is accessible by car. One road bisects the park, taking visitors along grand views, though there is nothing to the north and south, where the biggest glaciers and terrain exist. I was especially attentive to road access, as engine failure would have been a life-threatening complication during a forced landing and after. Nonetheless, I found myself snaking around immense peaks with near vertical mountainsides unlike anything I had yet seen in the Rockies. It felt like another world, looking at steep canyons three and four thousand feet below, with clear evidence that in the ancient past, the entire place would have had massive glacial features.

It was still more interesting to see the Great Plains just to the east. With little in the way of foothills, the Rockies give way to an abrupt descent to the plains, which then carry on for over a thousand miles, only slowly changing to forested terrain in the vicinity of the Great Lakes. This massive expanse of level land makes room for cold Canadian air masses to come south and strong winds to blow, likely converging with such aggressive terrain features to create atmospheric conditions that lent in the past to the formation of glaciers. Even today, Glacier National Park at 6,500 feet receives comparable snowfall to peaks of 14,000 feet in elevation in Colorado.

While I had an incredibly satisfying and unforgettable experience chasing these glaciers, it was hard to fathom how fast they are melting, and predictions that they will be gone in a decade or so. Terrain is so vast and immense that it seems impossible, though one only needs to look at early-season massive fires in the park during the season I took these flights, as well as burn scars right to timberline, along with beetle kill to the south. Glacier National Park won't last long. I am glad I preserved part of it in photos for posterity.

LEFT: *Ipasha Glacier (front), Chaney Glacier (rear).*   **BELOW:** *Pumpelly Glacier (left), Blackfoot Glacier (right).*   **ABOVE:** *Mt. Wilbur with Ahern Glacier.*

ABOVE: *Agassiz Glacier.* **BELOW:** *Vulture Glacier.* **RIGHT:** *Weasel Collar Glacier.*

LEFT: *Rainbow Glacier discharge.*   BELOW: *Carter Glaciers.*   ABOVE: *Kintia Glacier.*

BELOW: *Blackfoot Glacier.*

RIGHT & ABOVE: *Sperry Glacier.*

LEFT: *Harrison Glacier discharge.* **BELOW:** *Old Sun Glacier.* **ABOVE:** *Harrison Glacier.*

Above: *Jackson Glacier.*

Right & Below: *Pumpelly Glacier.*

LEFT: *Grinnell Glacier.*   **BELOW:** *Swiftcurrent Glacier.*   **ABOVE:** *Glacier beneath Clements Mountain.*

ABOVE: *Rainbow Glacier.* **BELOW:** *Chaney Glacier.* **RIGHT:** *Unnamed glaciers beneath Ahern Glacier.*

LEFT: *Helen Lake & Ahern Glacier.*   **BELOW:** *Chaney Glacier.*   **ABOVE:** *Shepard Glacier.*

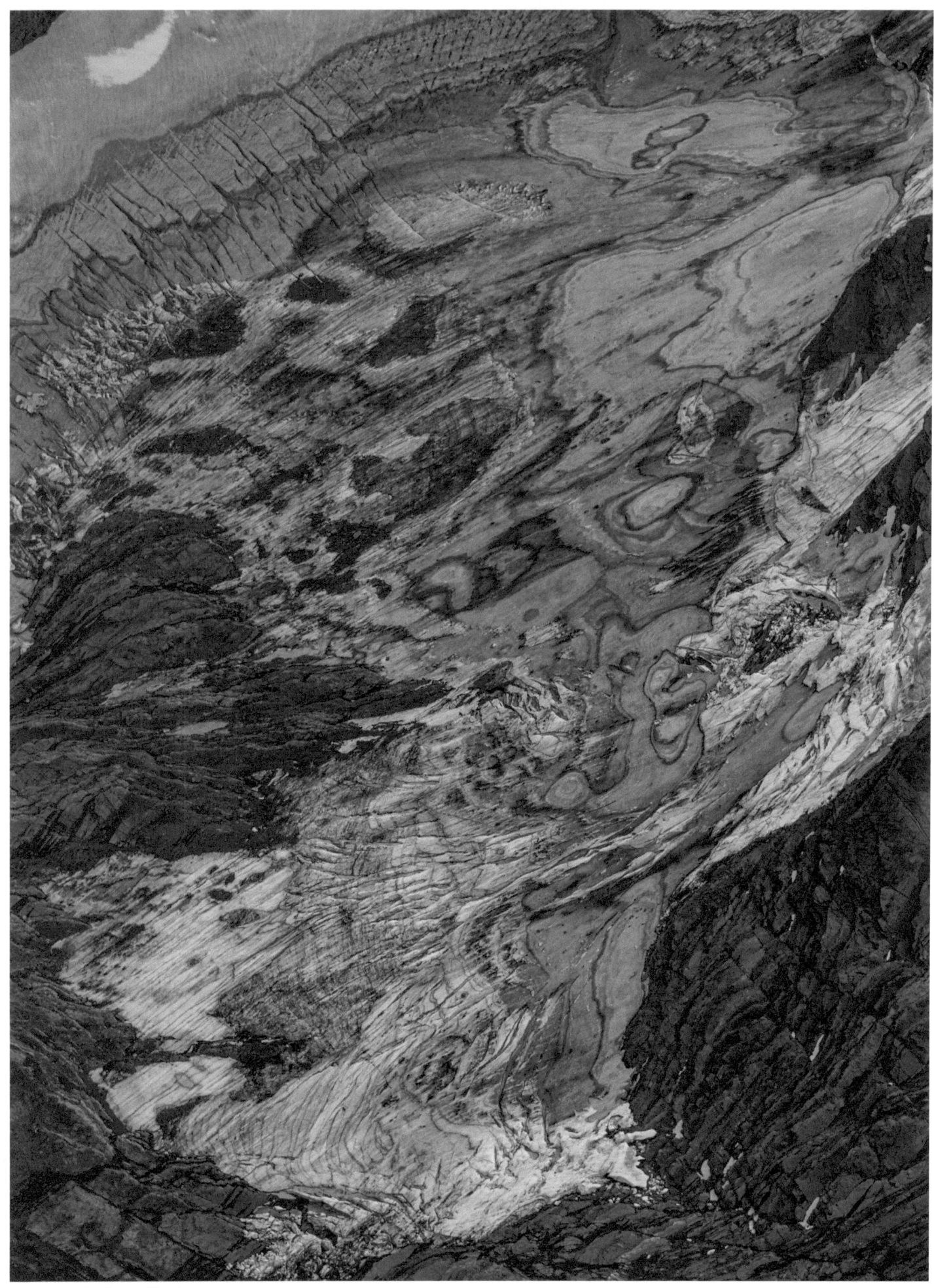

ABOVE: *Jackson Glacier.*

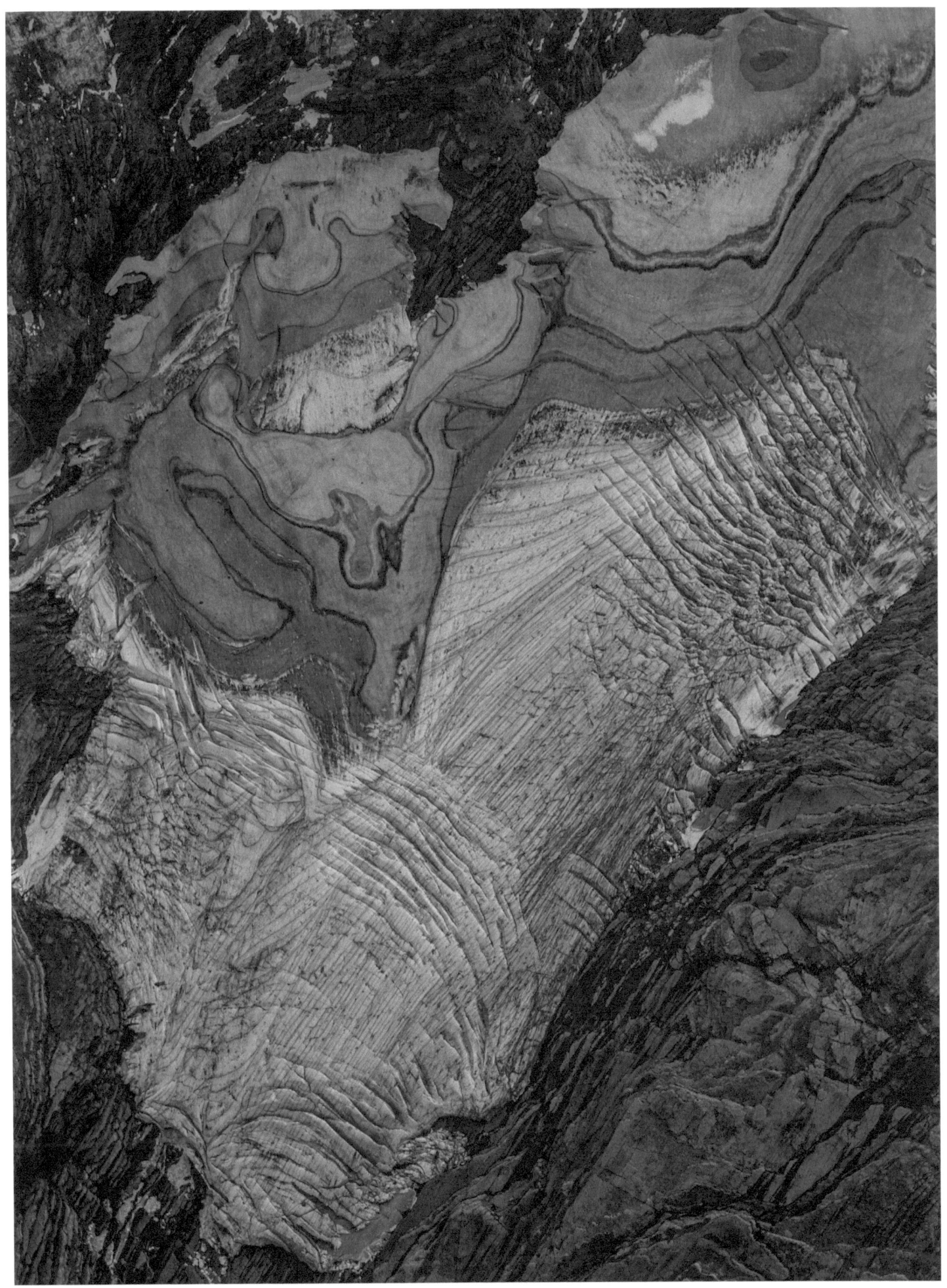
ABOVE: *Jackson Glacier.*

Above: *Two Ocean Glacier.*

Right & Below: *Pumpelly Glacier.*

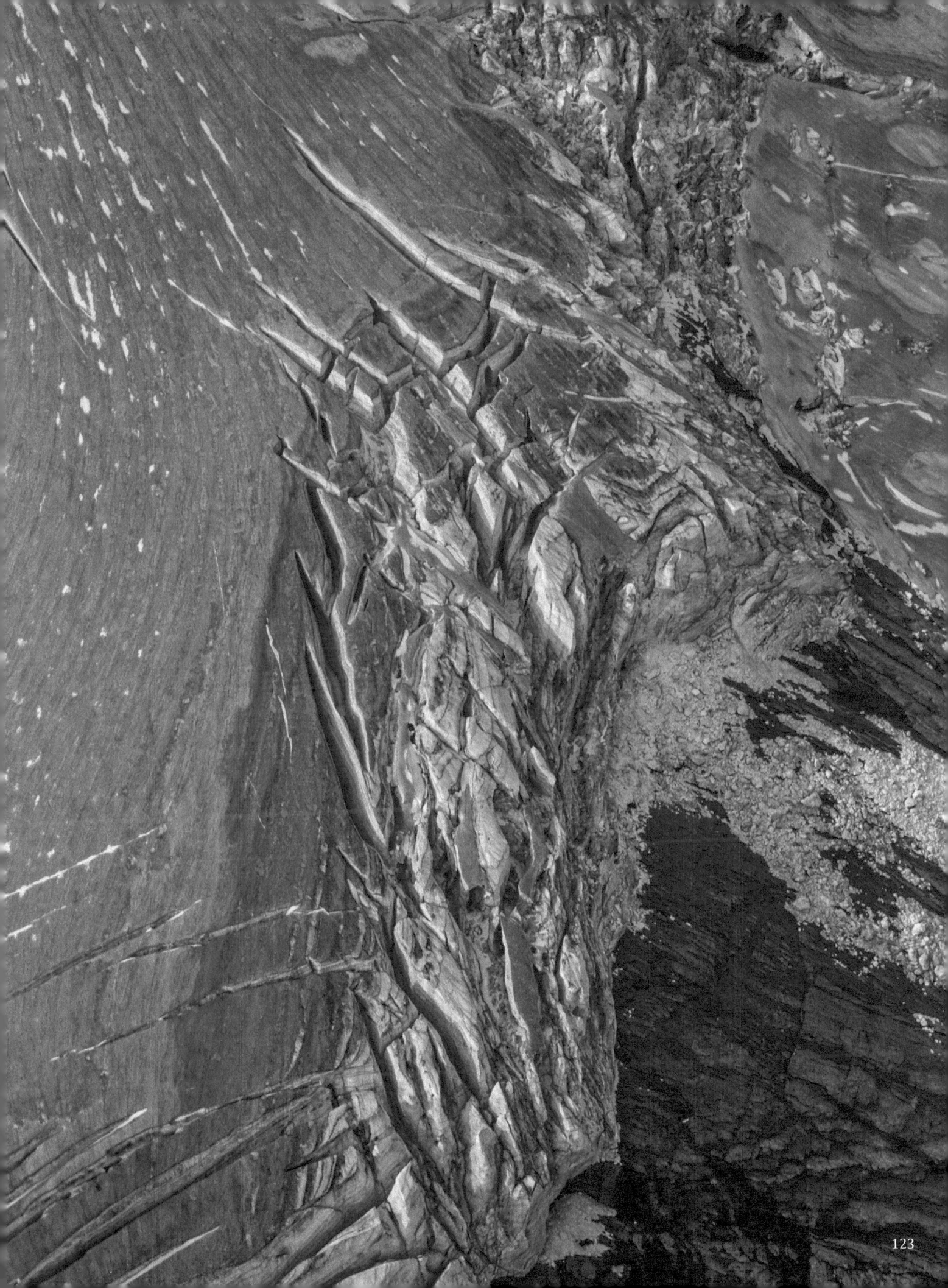

ABOVE: *Ahern Glacier.*

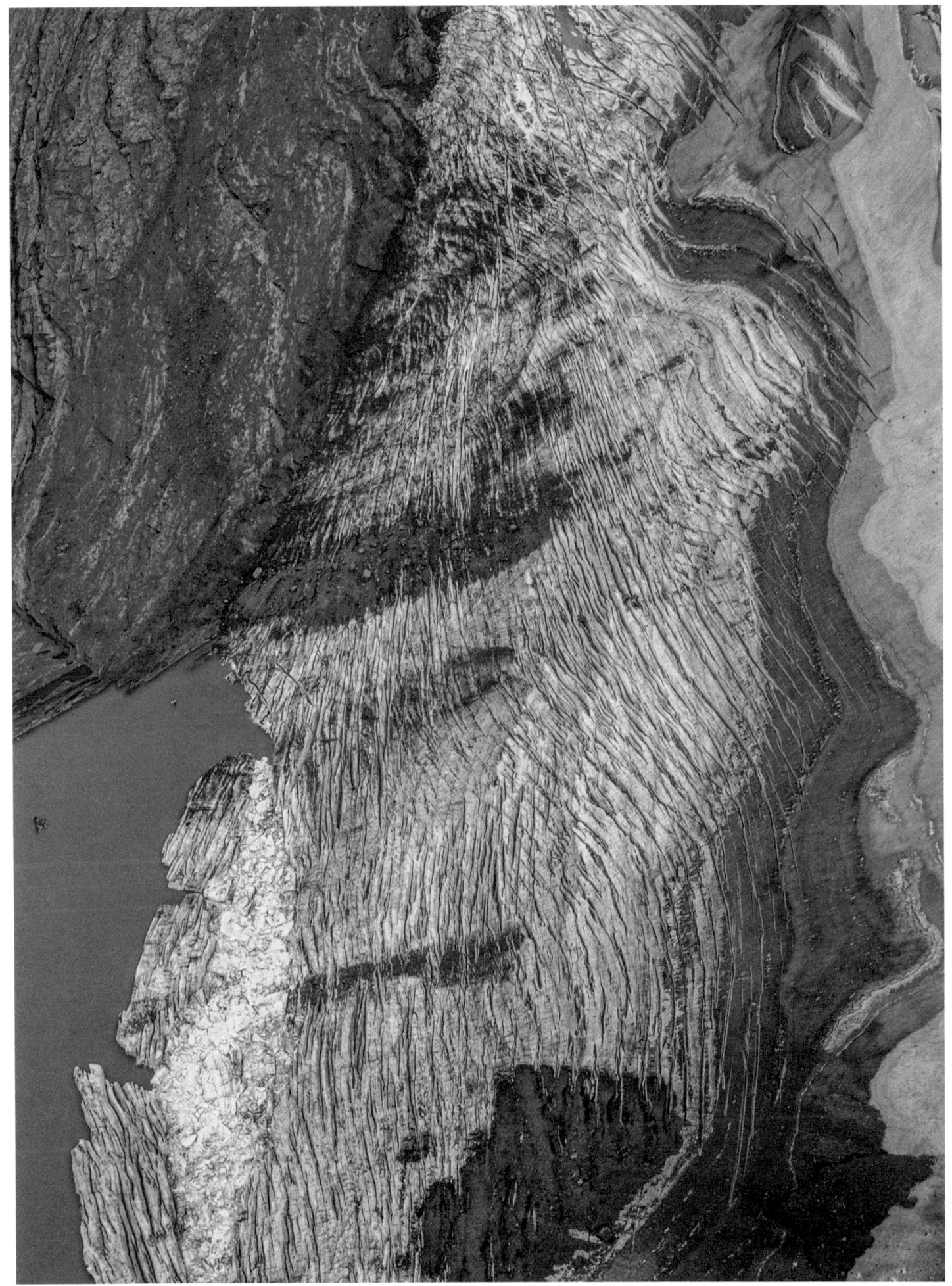

Above: *Grinnell Glacier.*

LEFT: *Two Ocean Glacier.* **BELOW:** *Kintia Glacier.* **ABOVE:** *Vulture Glacier.*

More Books by the Author

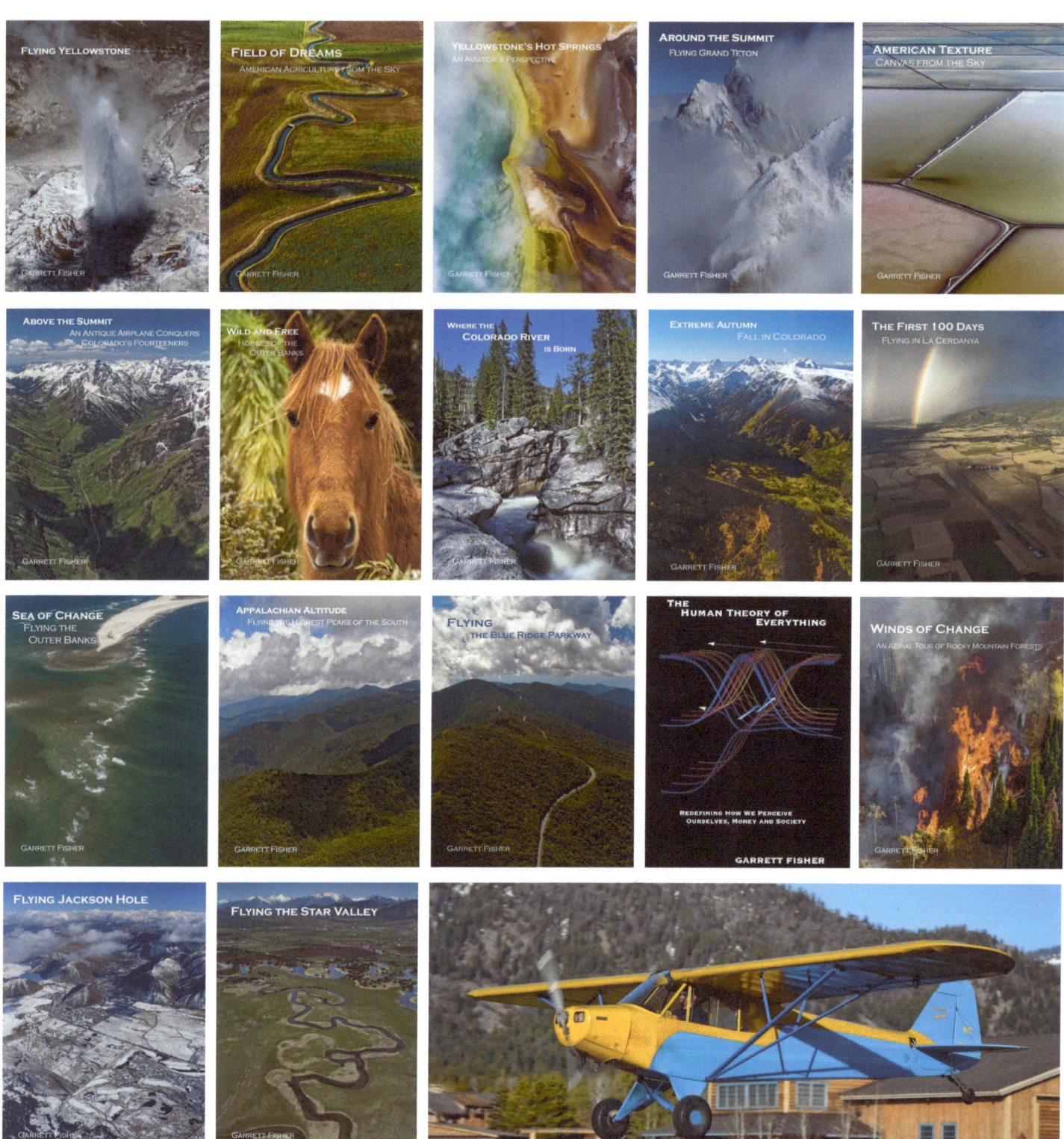

FOLLOW THE LATEST BOOK RELEASES AT:

WWW.GARRETTFISHER.ME

Photo: Adam Romer

Using the most ill-equipped aircraft possible for such adventure flying, Garrett Fisher has based his antique Piper PA-11 Cub Special in the Outer Banks of North Carolina, the highest airport in North America in Colorado, an elitist airpark near Yellowstone, the busiest airport in Germany, and now a secluded hideout in the Spanish Pyrenees. In the process, he has flown to an exhaustive list of rugged and dangerous places in America as well as some of the most beautiful sites in Europe, amassing an enormous collection of aerial photographs. Perpetually clueless about what is coming next, he continues wandering in the airplane, blogging about his adventures at www.garrettfisher.me.

www.ingramcontent.com/pod-product-compliance
Lightning Source LLC
Chambersburg PA
CBHW041930240526
45473CB00034B/722